2

ENERGY POLICIES, POLITICS AND PRICES

ENERGY POLICIES

ISSUES AND LEGISLATION

ENERGY POLICIES, POLITICS AND PRICES

Additional books and e-books in this series can be found on Nova's website under the Series tab.

ENERGY POLICIES, POLITICS AND PRICES

ENERGY POLICIES

ISSUES AND LEGISLATION

DOUG HIGGINS
EDITOR

snova
New York

Library of Congress Cataloging-in-Publication Data

ISBN: 978-1-53614-485-7

Published by Nova Science Publishers, Inc. † New York

CONTENTS

PREFACE

This book is a compilation of CRS reports on energy policies. Some topics discussed herein include the energy-related research and development of coal-based synthetic petroleum and atomic bombs, oil- and gas-related wells are, the Atomic Energy Act, the Federal Energy Regulatory Commission, energy tax provisions, the Clean Power Plan, Section 201 safeguard actions and the Low Income Home Energy Assistance Program.

Chapter 1 - Energy-related research and development (R&D)—on coal-based synthetic petroleum and on atomic bombs—played an important role in the successful outcome of World War II. In the postwar era, the federal government conducted R&D on fossil and nuclear energy sources to support peacetime economic growth. The energy crises of the 1970s spurred the government to broaden the focus to include renewable energy and energy efficiency. Over the 41-year period from the Department of Energy's (DOE's) inception at the beginning of fiscal year (FY) 1978 through FY2018, federal funding for renewable energy R&D amounted to about 18% of the energy R&D total, compared with 6% for electric systems, 16% for energy efficiency, 24% for fossil, and 37% for nuclear. For the 71-year period from 1948 through 2018, nearly 13% went to renewables, compared with nearly 5% for electric systems, 11% for energy efficiency, 24% for fossil, and 48% for nuclear.

Chapter 2 - Oil- and gas-related wells are documented in 110 (approximately 18%) of the 605 units of the National Wildlife Refuge System (NWRS). The U.S. Fish and Wildlife Service (FWS), in the Department of the Interior (DOI), administers the NWRS, which includes primarily national wildlife refuges, along with wetland management districts and waterfowl production areas. The wells in the NWRS most commonly involve nonfederal oil and gas resources but sometimes encompass federal resources. Oil and gas development in the NWRS has the potential to adversely impact wildlife and/or the environment, and some see it as contrary to the mission and purposes for which the NWRS was established. Others think that some levels of oil and gas activity may take place in refuges without harming the system's central mission of wildlife conservation and that such activity could benefit the U.S. economy and provide greater energy security. FWS, which administers nonfederal mineral activities on refuge lands, and the Bureau of Land Management (BLM), which administers federal mineral activities on refuge lands, have developed regulations that seek to minimize the adverse impacts of oil and gas development in the NWRS, among other purposes. Nonfederal oil and gas activities in refuges most often occur where FWS has acquired surface rights to refuge lands without acquiring mineral rights. In these cases, the entity (such as an individual, corporation, or tribe) that retains a valid existing right to the mineral estate has the right to develop the oil and gas resources pursuant to regulations established by FWS. According to FWS data, there are 107 NWRS units with nonfederal wells, and 45 of these units have active wells. Nonfederal oil and gas activities in the NWRS outside of Alaska are governed by a final rule promulgated by FWS on November 2016, "Management of Non-Federal Oil and Gas Rights." In contrast to these nonfederal activities, leasing and development of federal oil and gas resources within the NWRS generally is prohibited. The primary exception is when federal oil and gas leases predate the establishment or expansion of an NWRS unit, in which case the lease can be allowed to continue. According to BLM records, outside of Alaska, there are 11 NWRS units with federal oil and gas wells, all of which have at least 1 producing well. BLM regulations require FWS concurrence as to the time, place, and nature of oil and gas activities in refuges, in order to maximize protection for wildlife populations and habitat.

For both federal and nonfederal wells in the NWRS, regulation within Alaska is different from that in the rest of the United States. In addition to general FWS and BLM regulations, these units also are subject to requirements of Alaska-specific laws, including the Alaska National Interest Lands Conservation Act (ANILCA; P.L. 96-487). Within Alaska, Kenai National Wildlife Refuge has both federal and nonfederal oil and gas wells. Three other Alaskan units have nonfederal wells. Congress has debated both the extent of oil and gas activities in the NWRS and the compatibility of these activities with the NWRS's mission and purposes. One issue that has been debated for many years is whether to allow energy development in the Arctic National Wildlife Refuge in northeastern Alaska. P.L. 115-97, enacted in December 2017, established a federal oil and gas program for a portion of the refuge. Congress may continue to pursue oversight or legislation related to the implementation of this program, including issues related to limits on the footprint of development, compliance with the National Environmental Policy Act (42 U.S.C. §§4321-4347), or judicial review of legal challenges, among other matters. Congress also has addressed FWS's 2016 nonfederal oil and gas rule through both oversight and legislation, and it may continue to consider aspects of these regulations as well as the appropriate role for FWS in overseeing nonfederal oil and gas wells in the NWRS.

Chapter 3 - In order for the United States to engage in significant civilian nuclear cooperation with other states, it must conclude a framework agreement that meets specific requirements under Section 123 of the Atomic Energy Act (AEA). Significant nuclear cooperation includes the export of reactors, critical parts of reactors, and reactor fuel. The AEA also provides for export control licensing procedures and criteria for terminating cooperation. Congressional review is required for Section 123 agreements; the AEA establishes special parliamentary procedures by which Congress may act on a proposed agreement.

Chapter 4 - A 2013 rifle attack on a critical electric power substation in Metcalf, CA, marked a turning point for the U.S. electric power sector. The attack prompted utilities across the country to reevaluate and restructure their physical security programs. It also set in motion proceedings in Congress and at the Federal Energy Regulatory Commission (FERC) which resulted in a new mandatory *Physical Security Reliability Standard* (CIP-014) for bulk power asset owners promulgated by the North American Electric Reliability Corporation (NERC) in 2015. In the three years since FERC approved this new standard, security risks to the power grid have become an even greater concern in the electric utility industry. Reflecting these ongoing security concerns, legislative proposals in the 115th Congress include provisions directed at power grid physical security. Congress also continues its oversight of grid security and implementation of NERC's security standards. Three entities play key roles in standards oversight and support of implementation for bulk power physical security. NERC and FERC oversee implementation of the CIP-014 standards, while the Department of Energy plays a supporting role in helping bulk power asset owners to protect their critical infrastructure. The detailed findings of NERC's compliance activities are not publicly disclosed due to their confidential nature. However, NERC has stated that the utility industry is making progress towards effective implementation of the CIP-014 standard and NERC has been "encouraged" by grid security measures put in place so far. NERC compliance audits as of February 2018 have uncovered no major failures to date. In addition to compliance with NERC's standards, there have been other observable changes within the electricity sector reflecting greater emphasis on bulk power physical security. These changes include realignment in corporate structure to support physical security, incorporating physical security in transmission planning, new security products and services, utility capital investment in physical security, and utility participation in voluntary security programs. While public information about such changes is limited, it suggests they may be significant and widespread. Although the electric power sector seems to be moving in the overall direction of greater physical security for critical assets, many measures have yet to be implemented and the process of corporate realignment around physical security is still underway. NERC's CIP-014 standards have been promulgated recently, and bulk power asset owners have largely begun enhancing physical security under the standard over the last two years. Therefore, although it is probably accurate to conclude that, based on the objectives of the CIP-014 standards, the U.S. electric grid is more physically secure than it was five years ago, it has not necessarily reached the level of physical security needed based on the sector's own assessments of risk. Bulk power security remains a work in progress. Congress continues to be concerned about the current state of electric grid physical security. Among many specific issues of potential interest, Congress may focus on several with policy significance: security implementation oversight, cost recovery, hardening vs. resilience, and the quality of threat information. As CIP-014 implementation and other physical security initiatives proceed, Congress also may seek to maintain its focus on the power sector's overall progress, not only on short term compliance with NERC's security standards, but also on structural changes supporting physical security as a priority far into the future.

Chapter 5 - Thirteen temporary energy tax provisions expired at the end of 2017. All of these provisions had expired at the end of 2016, and were retroactively extended by the Bipartisan Budget Act of 2018 (BBA; P.L. 115-123) and made available for the 2017 tax year. This chapter briefly summarizes and discusses the economic impact of energy-related tax provisions that expired at the end of 2017, including the following:

- *Renewable energy property provisions:*
  - Production Tax Credit (PTC) for Non-Wind Facilities
- *Alternative and renewable fuels provisions:*
  - Incentives for Biodiesel and Renewable Diesel
  - Incentives for Alternative Fuel and Alternative Fuel Mixtures
  - Alternative Fuel Vehicle Refueling Property
  - Second Generation (Cellulosic) Biofuel Producer Credit
  - Special Depreciation Allowance for Second Generation (Cellulosic) Biofuel Plant Property
- *Vehicles provisions:*
  - Alternative motor vehicle credit for qualified fuel cell vehicles
  - Credit for two-wheeled plug-in electric vehicles

- *Building energy efficiency provisions:*
    - Credit for Construction of Energy-Efficient New Homes
    - Energy-Efficient Commercial Building Deduction
    - Credit for Section 25C Nonbusiness Energy Property
- *Other provisions:*
    - Special Rule to Implement Electric Transmission Restructuring
    - Credit for Production of Indian Coal

This chapter does not include provisions that in the past have been classified as individual or business related.

Chapter 6 - In 2015, when the U.S. Environmental Protection Agency (EPA) promulgated the Clean Power Plan to reduce greenhouse gas emissions from fossil-fueled electric power plants, it concluded that the benefits of reducing emissions would outweigh the costs by a substantial margin under the scenarios analyzed. EPA estimated benefits ranging from $31 billion to $54 billion in 2030 and costs ranging from $5.1 billion to $8.4 billion in 2030, when the rule would be fully implemented. In proposing to repeal the rule in October 2017, EPA revised the estimates of both its benefits and costs, finding in most cases that the benefits of the proposed repeal would outweigh the costs of the proposed repeal. However, EPA found that under other assumptions, the costs of the proposed repeal would outweigh the benefits of the proposed repeal. This report examines the changes in EPA's methodology that led to the revised conclusions about how benefits compare to costs. Three changes to the benefits estimates of the proposed repeal drive the agency's new conclusions.

- First, it considered only domestic benefits of the Clean Power Plan in its main analysis, excluding benefits that occur outside the United States.
- Second, it used different discount rates, including one higher rate, than the 2015 analysis to state the present value of future climate benefits expected from the Clean Power Plan.

- Third, the analysis reduced some estimates of the human health "co-benefits"— that is, the benefits resulting from pollutant reductions not directly targeted by the Clean Power Plan. Specifically, several scenarios assumed no health benefits below specified thresholds for some air pollutants.

EPA also changed the accounting treatment of demand-side energy efficiency savings. EPA's 2015 analysis treated savings from energy efficiency measures as a negative cost, whereas the 2017 analysis treated them as a benefit. Using the terminology of the proposed repeal, EPA moved energy savings from the cost savings estimate to the forgone benefits estimate. There was no change in the difference between benefits and costs because the benefits and costs increased by the same amount. This change took on more significance in a separate analysis that EPA conducted to analyze the cost savings of the proposed repeal. EPA based one set of benefit-cost estimates of the proposed repeal on its 2015 power sector modeling, which does not reflect changes that have since occurred in the power sector. EPA based the other set of benefit-cost estimates on more recent power sector projections from the Annual Energy Outlook 2017. The power sector changes subsequent to 2015 are potentially important and include changes in expected electricity demand, expected growth in electricity generation by renewable energy technologies, retirements of older generating units, changes in the prices and availability of different fuels and renewables, and state and federal regulations. While modeling differences render the two sets of estimates incomparable, both sets of estimates show a range of costs exceeding benefits (i.e., net costs), and benefits exceeding costs (i.e., net benefits) of the proposed repeal. EPA stated that it plans to update the power sector modeling and make it available for public comment before it finalizes the proposed repeal. This forthcoming analysis may show the extent to which updated power sector projections may change EPA's benefit-cost estimates.

Chapter 7 - On January 23, 2018, President Trump announced that he would impose additional tariffs on imports of_*large residential washing machines and solar photovoltaic (PV) cells and modules,* effective February 7, 2018. The President acted based on findings by the

International Trade Commission (ITC) that increased U.S. imports of these products were a "substantial cause of serious injury" to U.S. manufacturers, as a result of investigations under *Section 201* of the Trade Act of 1974 *(19 U.S.C. §2251,* as amended). When initiating the actions on January 23, the *President said,* "My administration is committed to defending American companies, and they've been very badly hurt from harmful import surges that threaten the livelihood of their workers, of jobs, actually, all over this country." While such actions may benefit some U.S. domestic producers, they could also raise prices for U.S. consumers and domestic industries that use these imports to manufacture downstream products. The Section 201 measures could also increase tensions with various U.S. trading partners. Section 201 "safeguard" actions provide temporary import relief (e.g., additional tariffs or quotas on imports) in order to facilitate positive adjustment of a U.S. industry to import competition. Safeguard investigations are unlike other trade remedy tools, such as *antidumping (AD)* and *countervailing duty (CVD)* cases that investigate "material injury" (or threat thereof) based on sales of imported products at less than fair value (AD) or that are subsidized by a foreign government or other public entity (CVD). Rather, Section 201 cases investigate import surges of fairly traded goods. Actions taken after Section 201, AD, and CVD investigations will be considered consistent with U.S. international trade obligations if they conform to the relevant *World Trade Organization (WTO) Agreements.* Other WTO members may use these agreements to challenge the United States' actions through the WTO's *dispute settlement process.*

Chapter 8 - The Energy Information Administration (EIA), in its *Short-Term Energy and Winter Fuels Outlook* (STEWFO) for the 2017-2018 winter heating season, projects that American consumers should expect to see heating expenditures that will be higher than last winter. However, the winter of 2016-2017 was relatively warm. Average expenditures for those heating with natural gas are projected to increase by 12%, while those heating with electricity are projected to see an increase of about 8%. These two fuels serve as the heating source for about 87% of all U.S. household heating. Propane and home heating oil consumers are also projected to see increased costs. Within the United States, average expenditures projections differences exist with respect to region of the country. Differences in weather conditions and fuel prices contribute to differing regional expenditures. Economic conditions of relatively high growth and low unemployment suggest that higher consumption levels of all fuels may occur. Increased consumption could lead to higher prices for all winter fuels. The STEWFO provides analysis of scenarios covering a warmer, or colder, winter than the base case forecast. The key federal program designed to assist low-income households is the Low Income Home Energy Assistance Program (LIHEAP). LIHEAP funding beyond the expiration of the Continuing Appropriations Resolution (P.L. 115-56) is uncertain and if reduced could put an additional burden on families qualified for benefits. If the weather in the winter of 2017-2018 is severe, or if energy prices are volatile, the fixed amount of LIHEAP funding will likely result in smaller assistance payments to families, or fewer families receiving payments.

In: Energy Policies: Issues and Legislation ISBN: 978-1-53614-485-7
Editor: Doug Higgins

Chapter 1

RENEWABLE ENERGY R&D FUNDING HISTORY: A COMPARISON WITH FUNDING FOR NUCLEAR ENERGY, FOSSIL ENERGY, ENERGY EFFICIENCY, AND ELECTRIC SYSTEMS R&D*

Corrie E. Clark

SUMMARY

Energy-related research and development (R&D)—on coal-based synthetic petroleum and on atomic bombs—played an important role in the successful outcome of World War II. In the postwar era, the federal government conducted R&D on fossil and nuclear energy sources to support peacetime economic growth. The energy crises of the 1970s spurred the government to broaden the focus to include renewable energy and energy efficiency. Over the 41-year period from the Department of Energy's (DOE's) inception at the beginning of fiscal year (FY) 1978

* This is an edited, reformatted and augmented version of a Congressional Research Service report, RS22858, dated June 18, 2018.

through FY2018, federal funding for renewable energy R&D amounted to about 18% of the energy R&D total, compared with 6% for electric systems, 16% for energy efficiency, 24% for fossil, and 37% for nuclear. For the 71-year period from 1948 through 2018, nearly 13% went to renewables, compared with nearly 5% for electric systems, 11% for energy efficiency, 24% for fossil, and 48% for nuclear.

INTRODUCTION

Cumulative Funding History

This chapter provides a cumulative history of Department of Energy (DOE) funding for renewable energy compared with funding for the other energy technologies—nuclear energy, fossil energy, energy efficiency, and electric systems. Specifically, it provides a comparison that covers cumulative funding over the past 10 years (FY2009-FY2018), a second comparison that covers the 41-year period since DOE was established at the beginning of FY1978 through FY2018, and a third comparison that covers a 71-year funding history (FY1948-FY2018) for DOE and predecessor agencies.

FY2017 and FY2018 Appropriations

The final amount of FY2017 Energy and Water Development appropriations for DOE energy technologies was established by the Consolidated Appropriations Act, 2017 (P.L. 115-31), which was signed by the President on May 5, 2017. The act contained appropriations for all FY2017 appropriations bills, including Energy and Water Development programs (Division D).[1]

Final funding for FY2018 was set by the Consolidated and Further Continuing Appropriations Act, 2018 (P.L. 115-141), which was signed by

[1] For more details, see CRS Report R44465, *Energy and Water Development: FY2017 Appropriations*, by Mark Holt.

the President on May 23, 2018. Funding levels for DOE are included in Energy and Water Development programs (Division D).[2]

Guide to Tables and Charts

Figure 1 presents the fiscal year funding totals for DOE in real terms (2016 dollars) since 1978 for each technology or energy source.

Table 1 shows the cumulative funding totals in real terms (2016 dollars) for the past 10 years (first column), 41 years (second column), and 71 years (third column). *Table 2* converts the data from *Table 1* into relative shares of spending for each technology or energy source, expressed as a percentage of total spending for each period.

Figure 2 displays the data from the first column of Table 2 as a pie chart. That chart shows the relative shares of cumulative DOE spending for each technology or energy source over the 10 years from FY2009 through FY2018. *Figure 3* provides a similar chart for the period from FY1978 through FY2018. *Figure 4* shows a chart for FY1948 through FY2018.

Background

Evolution of the Department of Energy

The availability of energy—especially gasoline and other liquid fuels—played a critical role in World War II.[3] Another energy-related factor was the application of research and development (R&D) to the atomic bomb (Manhattan Project) and other military technologies. During the post-World War II era, the federal government began to apply R&D to

---

[2] For more details, see CRS Report R44895, *Energy and Water Development: FY2018 Appropriations*, by Mark Holt and Corrie E. Clark.

[3] Regarding coal-based synthetic petroleum production—before, during, and after World War II—see the discussion by DOE's National Energy Technology Laboratory at https://www.netl.doe.gov/research/coal/energy-systems/gasification/gasifipedia/history-liquid-fuels.

the peacetime development of energy sources to support economic growth. At that time, the primary R&D focus was on fossil fuels and new forms of energy derived from nuclear fission and nuclear fusion.

The Atomic Energy Act of 1946 established the Atomic Energy Commission (AEC), which inherited all of the Manhattan Project's R&D activities and placed nuclear weapon development and nuclear power management under civilian control.[4] A major focus of the AEC was research on "atoms for peace," the use of nuclear energy for civilian electric power production.[5] Prompted by the oil embargo declared by the Organization of Arab Petroleum Exporting Countries in 1973, the Federal Energy Administration was established in mid-1974.[6] In early 1975, the Energy Research and Development Administration (ERDA) was established, incorporating the AEC and several energy programs that had been operating under the Department of the Interior and other federal agencies.[7]

The Department of Energy (DOE) was established by law in 1977,[8] incorporating activities of the FEA and ERDA. All of the energy R&D programs—fossil, nuclear, renewable, and energy efficiency—were brought under its administration. DOE also undertook a small program in energy storage and electricity system R&D that supports the four main energy technology programs.[9]

---

[4] Alice L. Buck, *A History of the Atomic Energy Commission*, DOE, July 1983, p. 1, https://nsarchive2.gwu.edu/nukevault/ebb507/docs/doc%2012%20%2083.07%20Buck%20AEC%20history.pdf.

[5] Ibid., p. 3.

[6] See Federal Energy Administration Act (P.L. 93-275). Terrence R. Fehner and Jack M. Holl, *Department of Energy 1977-1994: A Summary History*, DOE, DOE/HR-0098, November 1994, p. 17, https://www.energy.gov/management/downloads/fehner-and-holl-department-energy-1977-1994-summary-history.

[7] See Energy Reorganization Act of 1974 (P.L. 93-438). Terrence R. Fehner and Jack M. Holl, *Department of Energy 1977-1994: A Summary History*, DOE, DOE/HR-0098, November 1994, p. 18, https://www.energy.gov/management/downloads/fehner-and-holl-department-energy-1977-1994-summary-history.

[8] The Department of Energy Organization Act, P.L. 95-91, was enacted on August 4, 1977.

[9] This program includes R&D on advanced batteries to store electricity and transmission equipment to transfer electricity with less heat loss (i.e., at higher levels of energy efficiency).

Evolution of Energy Technology R&D Funding

From FY1948 through FY1977, the majority of federal government support for energy R&D focused on fossil energy and nuclear power technologies.[10] Total spending on fossil energy technologies over that period amounted to about $17.1 billion, in constant FY2016 dollars. The federal government spent about $51.6 billion (in constant FY2016 dollars) during that period for nuclear energy R&D (in all the tables and figures in this chapter, the "nuclear" category includes both fission and fusion).[11]

The energy crises of the 1970s spurred the federal government to expand its R&D programs to include renewable (wind, solar, biomass, geothermal, hydro) energy and energy efficiency technologies. Comparatively modest efforts to support renewable energy and energy efficiency began during the early 1970s.[12] Since FY1978, DOE has been the main supplier of energy R&D funding compared to other federal agencies.[13]

In real (constant dollar) terms, funding support for all four of the main energy technologies skyrocketed during the 1970s to a combined peak in FY1979 at about $8.6 billion (2016 constant dollars). Funding then dropped steadily, to about $2.0 billion (2016 dollars) per year during the late 1990s. Since then, funding has increased gradually—except that the American Recovery and Reinvestment Act of 2009 (P.L. 111-5) provided a one-year spike of $13 billion (2016 dollars) in FY2009. For FY2018, DOE energy R&D funding stood at nearly $4.5 billion (2016 dollars).

---

[10] DOE. Pacific Northwest Laboratory. *An Analysis of Federal Incentives Used to Stimulate Energy Production.* 1980. The spending for fossil energy included coal, oil, and natural gas technologies.

[11] Ibid.

[12] According to data compiled from DOE's *Conservation and Renewable Energy Base Table,* February 1990, from FY1973 through FY1977, the federal government spent about $1.7 billion (in constant FY2016 dollars) on renewable energy R&D, $413 million on energy efficiency R&D, and $246 million on electric systems R&D.

[13] There has been some energy R&D funding—on a much smaller scale—provided by the Department of Defense and other agencies. Coverage of that funding is beyond the scope of this report.

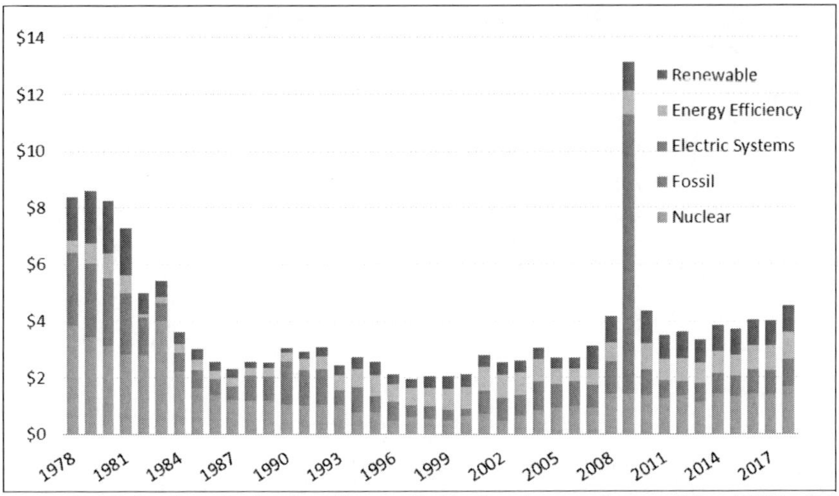

Source: *DOE Budget Authority History Table by Appropria*tion, May 2007; *DOE Congressional Budget Requests* (several years); Deflator Source: *The Budget for Fiscal Year 2019*. Historical Tables. Table 10.1. Gross Domestic Product and Deflators Used in the Historical Tables, 1940-2023.

Notes: Column for FY2009 includes funding provided in the American Recovery and Reinvestment Act of 2009 (P.L. 111-5).

Figure 1. Fiscal Year R&D Funding of DOE Energy Technologies (billions of 2016 dollars).

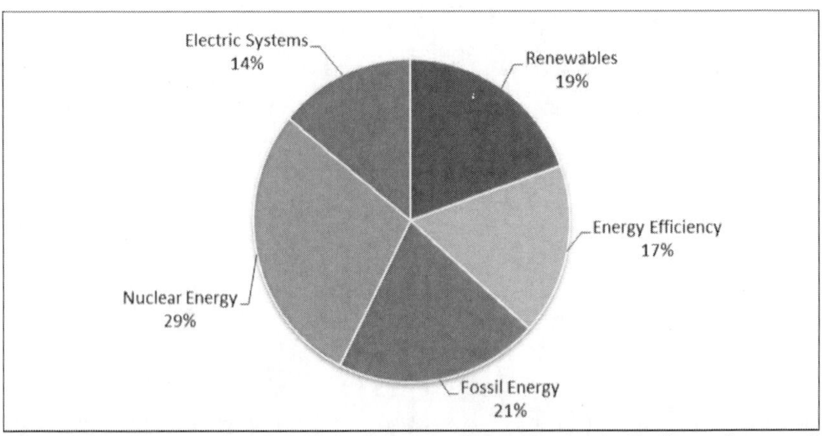

Source: *DOE Congressional Budget Requests* (several years); Deflator Source: *The Budget for Fiscal Year 2019*. Historical Tables. Table 10.1. Gross Domestic Product and Deflators Used in the Historical Tables, 1940-2023.

Figure 2. DOE Energy Technology Share of Funding, FY2009-FY2018.

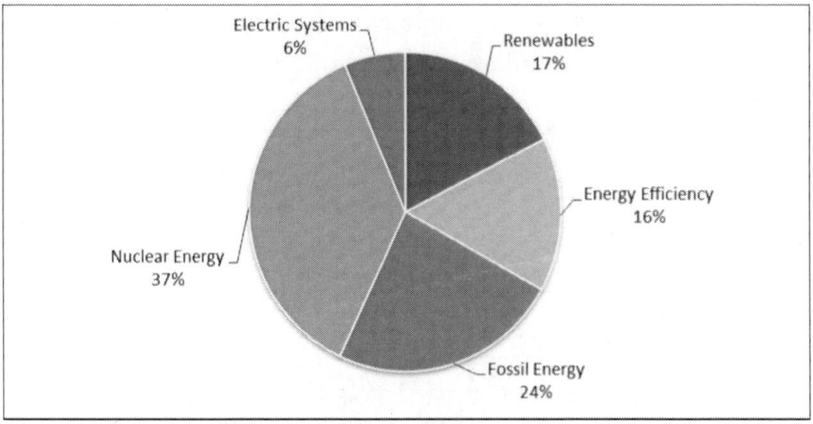

Sources: *DOE Budget Authority History Table by Appropriation*, May 2007; *DOE Congressional Budget Requests* (several years); Deflator Source: *The Budget for Fiscal Year 2019*. Historical Tables. Table 10.1. Gross Domestic Product and Deflators Used in the Historical Tables, 1940-2023.

Figure 3. DOE Energy Technology Share of Funding, FY1978-FY2018.

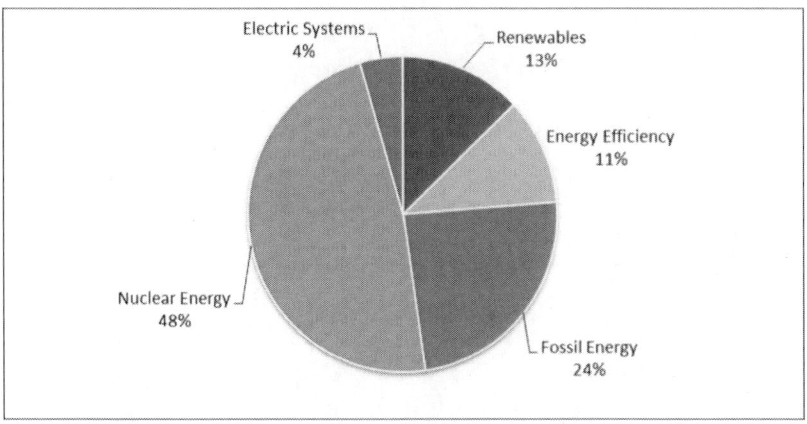

Sources: *DOE Budget Authority History Table by Appropriation*, May 2007; *DOE Congressional Budget Requests* (several years); DOE (Pacific Northwest Laboratory), *An Analysis of Federal Incentives Used to Stimulate Energy Production*, 1980; DOE Conservation and Renewable Energy Base Table. Feb. 1990. Deflator Source: *The Budget for Fiscal Year 2019*. Historical Tables. Table 10.1, Gross Domestic Product and Deflators Used in the Historical Tables, 1940-2023.

Figure 4. DOE Energy Technology Share of Funding, FY1948-FY2018.

Table 1. DOE Energy Technology Cumulative Funding Totals (billions of 2016 dollars)

Technology	Period FY2009-FY2018 (10 years)	FY1978-FY2018 (41 years)	FY1948-FY2018 (71 years)
Renewable Energy	$ 9.37	$ 27.65	$ 29.35
Energy Efficiency	8.22	24.73	25.14
Fossil Energy	9.98	37.85	54.96
Nuclear Energy	13.71	57.99	109.59
Electric Systems	6.74	10.06	10.31
Total	$ 48.01	$ 158.28	$ 229.35

Sources: *DOE Budget Authority History Table by Appropriation*, May 2007; *DOE Congressional Budget Requests* (several years); DOE (Pacific Northwest Laboratory), *An Analysis of Federal Incentives Used to Stimulate Energy Production*, 1980. Deflator Source: *The Budget for Fiscal Year 2019*. Historical Tables. Table 10.1. Gross Domestic Product and Deflators Used in the Historical Tables, 1940-2023.

Table 2. DOE Energy Technology Share of Funding (percentage; derived from Table 1)

Technology	Period FY2009-FY2018 (10 years)	FY1978-FY2018 (41 years)	FY1948-FY2018 (71 years)
Renewable Energy	19.5%	17.5%	12.8%
Energy Efficiency	17.1%	15.6%	11.0%
Fossil Energy	20.8%	23.9%	24.0%
Nuclear Energy	28.6%	36.6%	47.8%
Electric Systems	14.0%	6.4%	4.5%
Total	100.0%	100.0%	100.0%

Sources: *DOE Budget Authority History Table by Appropriation*, May 2007; *DOE Congressional Budget Requests* (several years); DOE (Pacific Northwest Laboratory), *An Analysis of Federal Incentives Used to Stimulate Energy Production*, 1980; DOE Conservation and Renewable Energy Base Table. February 1990. Deflator Source: *The Budget for Fiscal Year 2019*. Historical Tables. Table 10.1. Gross Domestic Product and Deflators Used in the Historical Tables, 1940-2023.

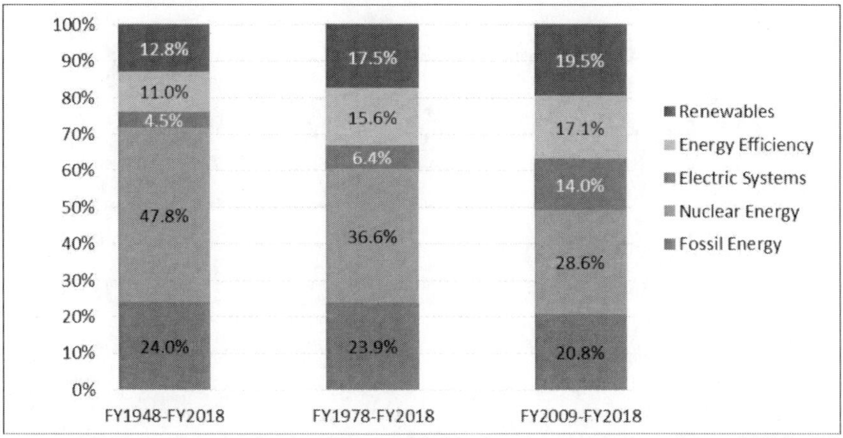

Source: *DOE Budget Authority History Table by Appropriation*, May 2007; *DOE Congressional Budget Requests* (several years); DOE (Pacific Northwest Laboratory), *An Analysis of Federal Incentives Used to Stimulate Energy Production*, 1980; *DOE Conservation and Renewable Energy Base Table*. Feb. 1990. Deflator Source: *The Budget for Fiscal Year 2019*. Historical Tables. Table 10.1, Gross Domestic Product and Deflators Used in the Historical Tables, 1940-2023.

Notes: Column to far left shows shares for the period FY1948-FY2018; middle column shows shares for period from FY1978-FY2018; and far right column shows shares for period from FY2009-FY2018.

Figure 5. DOE Energy Technology Share of Funding, Comparison over Three Periods.

In: Energy Policies: Issues and Legislation ISBN: 978-1-53614-485-7
Editor: Doug Higgins

Chapter 2

OIL AND GAS ACTIVITIES WITHIN THE NATIONAL WILDLIFE REFUGE SYSTEM*

R. Eliot Crafton, Laura B. Comay and Marc Humphries

SUMMARY

Oil- and gas-related wells are documented in 110 (approximately 18%) of the 605 units of the National Wildlife Refuge System (NWRS). The U.S. Fish and Wildlife Service (FWS), in the Department of the Interior (DOI), administers the NWRS, which includes primarily national wildlife refuges, along with wetland management districts and waterfowl production areas. The wells in the NWRS most commonly involve nonfederal oil and gas resources but sometimes encompass federal resources. Oil and gas development in the NWRS has the potential to adversely impact wildlife and/or the environment, and some see it as contrary to the mission and purposes for which the NWRS was established. Others think that some levels of oil and gas activity may take place in refuges without harming the system's central mission of wildlife conservation and that such activity could benefit the U.S. economy and provide greater energy security. FWS, which administers nonfederal mineral activities on refuge lands, and the Bureau of Land Management

* This is an edited, reformatted and augmented version of a Congressional Research Service report, R45192, dated May 9, 2018.

(BLM), which administers federal mineral activities on refuge lands, have developed regulations that seek to minimize the adverse impacts of oil and gas development in the NWRS, among other purposes.

Nonfederal oil and gas activities in refuges most often occur where FWS has acquired surface rights to refuge lands without acquiring mineral rights. In these cases, the entity (such as an individual, corporation, or tribe) that retains a valid existing right to the mineral estate has the right to develop the oil and gas resources pursuant to regulations established by FWS. According to FWS data, there are 107 NWRS units with nonfederal wells, and 45 of these units have active wells. Nonfederal oil and gas activities in the NWRS outside of Alaska are governed by a final rule promulgated by FWS on November 2016, "Management of Non-Federal Oil and Gas Rights."

In contrast to these nonfederal activities, leasing and development of federal oil and gas resources within the NWRS generally is prohibited. The primary exception is when federal oil and gas leases predate the establishment or expansion of an NWRS unit, in which case the lease can be allowed to continue. According to BLM records, outside of Alaska, there are 11 NWRS units with federal oil and gas wells, all of which have at least 1 producing well. BLM regulations require FWS concurrence as to the time, place, and nature of oil and gas activities in refuges, in order to maximize protection for wildlife populations and habitat.

For both federal and nonfederal wells in the NWRS, regulation within Alaska is different from that in the rest of the United States. In addition to general FWS and BLM regulations, these units also are subject to requirements of Alaska-specific laws, including the Alaska National Interest Lands Conservation Act (ANILCA; P.L. 96-487). Within Alaska, Kenai National Wildlife Refuge has both federal and nonfederal oil and gas wells. Three other Alaskan units have nonfederal wells.

Congress has debated both the extent of oil and gas activities in the NWRS and the compatibility of these activities with the NWRS's mission and purposes. One issue that has been debated for many years is whether to allow energy development in the Arctic National Wildlife Refuge in northeastern Alaska. P.L. 115-97, enacted in December 2017, established a federal oil and gas program for a portion of the refuge. Congress may continue to pursue oversight or legislation related to the implementation of this program, including issues related to limits on the footprint of development, compliance with the National Environmental Policy Act (42 U.S.C. §§4321-4347), or judicial review of legal challenges, among other matters. Congress also has addressed FWS's 2016 nonfederal oil and gas rule through both oversight and legislation, and it may continue to consider aspects of these regulations as well as the appropriate role for FWS in overseeing nonfederal oil and gas wells in the NWRS.

INTRODUCTION

The development of oil and natural gas resources in the National Wildlife Refuge System (NWRS) has been the subject of administrative rulemaking and the focus of debate in Congress.[1] The NWRS is a network of lands and waters administered by the U.S. Fish and Wildlife Service (FWS), an agency within the Department of the Interior (DOI).[2] The NWRS contains 566 national wildlife refuges, 36 wetland management districts composed of waterfowl production areas (WPAs), and 3 WPAs that lay outside of wetland management districts.[3] Issues pertaining to oil and gas wells in the NWRS include private property rights, economic and energy security benefits of oil and gas development, and the potential for adverse effects of oil and gas development on NWRS lands and wildlife. Some see these activities as contrary to the mission of the NWRS as stated in the National Wildlife Refuge System Administration Act, as amended (NWRSAA; 16 U.S.C. §668dd), whereas others think that oil and gas activities can be managed so as to avoid undue harm to wildlife or that U.S. energy needs outweigh conservation concerns. Both the executive branch and Congress have addressed oil and gas activity in the NWRS through

- debating the compatibility of oil and gas activities with the purposes of the NWRS;

---

[1] For general information on oil and gas activities in the National Wildlife Refuge System (NWRS), see U.S. Fish and Wildlife Service (FWS), NWRS, "Oil and Gas," October 31, 2017, at https://www.fws.gov/refuges/oil-and-gas/, hereinafter cited as FWS, *NWRS Oil and Gas*.

[2] As defined in the National Wildlife Refuge System Administration Act (NWRSAA; 16 U.S.C. §668dd(a)), the mission of the NWRS is to "administer a national network of lands and waters for the conservation, management, and where appropriate, restoration of the fish, wildlife, and plant resources and their habitats within the United States for the benefit of present and future generations of Americans."

[3] The NWRS also includes national monument areas and coordination areas, which are not included in the total unit count within this report.

- examining the potential economic and energy security benefits of developing oil and gas resources in the NWRS, predominantly in Alaska; and
- promulgating regulations for oil and gas activities in the NWRS to ensure access to nonfederal mineral rights while attempting to minimize impacts on natural resources, including lands, waters, and wildlife.

FWS states that it aims to balance the rights of mineral resource owners and lessees in the NWRS with the system's conservation mission.[4] In 2016, FWS promulgated a final rule, titled "Management of Non-Federal Oil and Gas Rights," to update regulations for nonfederal oil and gas activities on NWRS land.[5] The 115th Congress also has considered the regulation of nonfederal oil and gas activities in the NWRS and has enacted legislation to establish a federal oil and gas program in the Arctic National Wildlife Refuge (ANWR) in Alaska.[6]

BACKGROUND

Wells associated with oil and gas activities are present in about 18% of NWRS units. Most oil and gas operations on refuge lands involve nonfederal resources. These activities most often occur when rights to the surface and subsurface estates have been severed and FWS has acquired only the surface rights for addition to the NWRS, but valid existing rights to the mineral estate remain in nonfederal ownership.[7] FWS regulation of

---

[4] For more information, see FWS, NWRS, "Oil and Gas."

[5] FWS, "Management of Non-Federal Oil and Gas Rights," 81 *Federal Register* 79948-79981, November 14, 2016.

[6] For more information on the Arctic National Wildlife Refuge (ANWR), see CRS In Focus IF10782, *Arctic National Wildlife Refuge (ANWR) Provisions in P.L. 115-97, Tax Cuts and Jobs Act*, by Laura B. Comay, and CRS Report RL33872, *Arctic National Wildlife Refuge (ANWR): An Overview*, by Laura B. Comay, Michael Ratner, and R. Eliot Crafton.

[7] Nonfederal oil and gas rights in Alaskan refuges also may have been legislatively conveyed under the Alaska Native Claims Settlement Act (ANCSA; 43 U.S.C. §§1601 et seq.). See

these nonfederal activities is based on statutory authorities reflected in a 2016 agency final rule and in sections of the FWS *Service Manual*.[8]

Leasing and development of federally owned resources within the NWRS typically is prohibited. However, in certain cases, such as where the federal leases predate the establishment or expansion of the refuge, there may be federal leases and development on refuge lands.[9] The administration of federal oil and gas resources in the NWRS is delegated to the Bureau of Land Management (BLM), in the DOI, under the Mineral Leasing Act of 1920 (30 U.S.C. §§181 et seq.). BLM administers federal oil and gas leasing under regulations at Title 43, Part 3100, of the *Code of Federal Regulations*. The regulations require BLM to obtain FWS concurrence as to the time, place, and nature of oil and gas operations, "in order to give complete protection to wildlife populations and wildlife habitat on the areas leased."[10]

Oil and gas activities in Alaskan NWRS units, like those in other states, are administered by FWS for nonfederal operations and by BLM for federal operations. However, Alaskan activities are managed pursuant to Alaska-specific law as well as the general statutory and regulatory framework described above. Relevant Alaska laws include the Alaska National Interest Lands Conservation Act (ANILCA; P.L. 96-487) and the Alaska Native Claims Settlement Act (ANCSA; 43 U.S.C. §§1601 et seq.), among others.

This chapter contains sections on recent developments related to oil and gas wells in the NWRS, nonfederal wells in NWRS units outside of Alaska, federal wells in NWRS units outside of Alaska, nonfederal and federal wells in Alaskan NWRS units, and issues for Congress in

---

section on Alaska refuges ("Oil and Gas Activities in Alaska in the National Wildlife Refuge System") for more information.

[8] FWS, "Management of Non-Federal Oil and Gas Rights," 81 *Federal Register* 79948-79981, November 14, 2016. FWS, *Service Manual, 612 FW 1, Series: Natural and Cultural Resources Management, Part 612: Mineral Management*, December 27, 2016, at https://www.fws.gov/policy/612fw1.pdf, outlines FWS policy for oil and gas programs and identifies the various statutory authorities. See 50 C.F.R. Part 29, Subpart D, "Management of Non-Federal Oil and Gas Rights."

[9] FWS, NWRS, "Oil and Gas."

[10] 43 C.F.R. 3101.5-1.

considering oil and gas activities in the NWRS. In the sections on federal and nonfederal wells, the chapter presents information on the:

- number and location of oil and gas wells;
- compatibility of oil and gas activities with NWRS purposes; and
- regulatory mechanisms for the administration of oil and gas activities.

RECENT DEVELOPMENTS

Arctic National Wildlife Refuge

On December 22, 2017, President Trump signed into law P.L. 115-97, which provides for the creation of an oil and gas program in a portion of the Arctic National Wildlife Refuge (ANWR, or the refuge) in northeastern Alaska.[11] The law's enactment came after a decades-long debate over whether to allow oil and gas development in ANWR. P.L. 115-97 establishes an oil and gas program for the refuge's Coastal Plain,[12] with at least two oil and gas lease sales (of no fewer than 400,000 acres each) required in the next 10 years, and contains provisions for the distribution of revenues and royalties. P.L. 115-97 limits surface development to 2,000 acres for production and support facilities, which need not be concentrated in a single area.[13] Development proponents contend that the oil and gas program will generate economic activity, contribute to U.S. energy

---

[11] For more information, see CRS In Focus IF10782, *Arctic National Wildlife Refuge (ANWR) Provisions in P.L. 115- 97, Tax Cuts and Jobs Act*, by Laura B. Comay, and CRS Report RL33872, *Arctic National Wildlife Refuge (ANWR): An Overview*, by Laura B. Comay, Michael Ratner, and R. Eliot Crafton.

[12] The Coastal Plain (also known as the 1002 area in reference to the Alaska National Interest Lands Conservation Act; ANILCA, P.L. 96-497, 16 U.S.C. §§3161-3173) is a 1.57 million acre portion in the northeast corner of the Arctic National Wildlife Refuge (ANWR, or the refuge).

[13] P.L. 115-97, §20001(c)(3), explicitly includes "airstrips and any area covered by gravel berms or piers for support of pipelines" in its description of production and support facilities. It is unclear what other types of facilities or surface disturbances may be included or excluded. For more information, see the CRS reports cited in footnote 11.

security, and result in royalty revenues for both the federal government and the state of Alaska; opponents contend that the region's reserves are not necessary for energy independence and that development will detrimentally impact the refuge's unique biological resources.

P.L. 115-97 requires the first ANWR lease sale within four years of the law's enactment—by December 2021. Activities preparatory to the lease sale include identifying lands to be leased, conducting sale-specific environmental reviews, issuing notices of sales, and other "prelease" activities. Activities also could include new geological and geophysical surveys to determine the extent and location of hydrocarbon resources.

FWS Rule on Nonfederal Oil and Gas Management

Regulations at Title 50, Part 29, Subparts C and D of the *Code of Federal Regulations* govern nonfederal oil and gas operations that occur within the NWRS outside of Alaska.[14] In November 2016, FWS promulgated a final rule, "Management of Non-Federal Oil and Gas Rights," which amended these regulations with regard to the management of nonfederal oil and gas wells in the NWRS.[15] According to FWS, the changes, which went into effect on December 14, 2016, aimed to increase the consistency of FWS regulations with state and federal law and with best management practices. The rule is discussed further below under "Nonfederal Oil and Gas Operations."

The rule has been the subject of both congressional attention and administrative action in the 115th Congress. H.J.Res. 45, introduced on January 30, 2017, would have disapproved the FWS final rule under the Congressional Review Act (CRA; 5 U.S.C. §§801-808) and prevented

---

[14] 50 C.F.R. §29. For the purpose of this report, *nonfederal oil and gas operations* refers to oil and gas wells and associated activities that occur within the boundaries of a unit of the NWRS but are operated by nonfederal (state or private) entities.

[15] FWS, "Management of Non-Federal Oil and Gas Rights," 81 *Federal Register* 79948-79981, November 14, 2016. This rule became effective on December 14, 2016, updating a previous rule.

FWS from enacting a substantially similar rule in the future.[16] No further actions were taken on H.J.Res. 45, and the deadline to vote using fast-track procedures in the Senate pursuant to the CRA has elapsed.[17] Some proponents of the CRA action to overturn the FWS rule stated that this action was necessary because FWS had overstepped its authority in promulgating the rule.[18] They also stated that this rule would increase costs for oil and gas operators developing nonfederal resources in the NWRS. Some opponents of the CRA action contended that the final rule is a much-needed update to the previous management rule, which was more than 50 years old when the new rule was promulgated, and that the final rule establishes appropriate responsibilities for operators.[19]

The Administration also has been active on NWRS oil and gas operations. On March 28, 2017, President Trump issued Executive Order (E.O.) 13783, which required a review of several oil- and gas-related rules, including the FWS 2016 rule.[20] E.O. 13783 established that "it is in the national interest to promote clean and safe development of our Nation's vast energy resources, while at the same time avoiding regulatory burdens that unnecessarily encumber energy production, constrain economic growth, and prevent job creation."[21]

E.O. 13783 instructed the Secretary of the Interior to review and "if appropriate ... suspend, revise, or rescind the guidance" of the specified rules, including the FWS nonfederal oil and gas rule, to be consistent with

---

[16] For more information on the Congressional Review Act (5 U.S.C. §§801-808), see CRS Report R43992, *The Congressional Review Act (CRA): Frequently Asked Questions*, by Maeve P. Carey and Christopher M. Davis.

[17] 5 U.S.C. §802.

[18] These proponents state that private property, including nonfederal mineral rights, within NWRS units already is regulated under state law and that the rule is unwarranted given that FWS already may require special-use permits to protect NWRS surface estates. For example, see Corbin Hiar, "Agencies to Review 2 Less-Controversial Drilling Rules," *E&E News*, April 25, 2017.

[19] For example, see NWRA, "Refuge Oil and Gas Rule and the Congressional Review Act," at https://www.refugeassociation.org/wp-content/uploads/2017/02/oil-and-gas-rule-fact-sheet-FINAL.pdf.

[20] Executive Order 13783, "Promoting Energy Independence and Economic Growth," 82 *Federal Register* 16093- 16097, March 31, 2017.

[21] Ibid, p. 16903.

the statement of national interest.[22] When DOI published its final report on E.O. 13783, FWS was still in the process of reviewing the final rule.[23]

NONFEDERAL OIL AND GAS OPERATIONS IN THE NATIONAL WILDLIFE REFUGE SYSTEM

The FWS "Management of Non-Federal Oil and Gas Rights" rule revised requirements applicable to nonfederal oil and gas activities that occur in the NWRS outside of Alaska.[24] This rule replaced prior regulations governing these activities, which, according to FWS, had "remained unchanged for more than 50 years and provide[d] only vague guidance to staff and operators."[25] FWS said the 2016 final rule aimed to update the agency's management practices and improve the compatibility of nonfederal oil and gas activities with the mission of the NWRS. The rule provides an explicit process for refuge staff to follow to minimize the impact of oil and gas activities on refuge resources to the extent practicable. The process includes permitting requirements, operating standards, requirements for financial bonding, and penalty provisions.[26]

The rule also requires operators to meet performance-based standards, including standards for surface use and site management, specific resource protections, spill prevention and response, waste management, and

---

[22] Ibid, p. 16096.

[23] Department of the Interior, Office of the Secretary, "Final Report: Review of the Department of the Interior Actions that Potentially Burden Domestic Energy," 82 *Federal Register* 50545, November 1, 2017.

[24] FWS, "Management of Non-Federal Oil and Gas Rights," 81 *Federal Register* 79948-79981, November 14, 2016.

[25] Ibid., p. 79948.

[26] Ibid. Under the rule, new operations are by permit only. Existing permit holders must abide by terms in the permit and all state, local, and federal laws. Current operations not under a service permit are deemed "pre-existing operators and may continue to operate as they have been" and in compliance with state, local, and federal laws. (The rule states: "Making violation of applicable State laws related to oil and gas a prohibited act under the rule allows the Service to enforce these requirements as Federal requirements, and so gives us greater enforcement capabilities in ensuring that unnecessary impacts from these operations, such as leaks and spills, are avoided or minimized.") All operators must obtain a permit for any new production activities or for plugging and reclamation activities.

reclamation.[27] FWS claims that the performance-based standards, rather than prescriptive regulations, are intended to provide operators with flexibility in meeting the requirements of the permit while ensuring that NWRS resources are protected. These standards also address resource protection needs in different environments by allowing resource managers and operators to identify the most appropriate management practices for a given situation.

FWS also outlines its minerals management policy in the FWS *Service Manual.*[28] Specifically, Section 1.7(B) of the manual states that owners of nonfederal mineral rights within NWRS units outside of Alaska can transfer or develop these rights but must do so using the "technologically feasible, least damaging methods."[29]

Number and Location of Nonfederal Oil and Gas Operations

Prior to the issuance of the 2016 final rule, FWS published an environmental impact statement (EIS) in August 2016. As part of the EIS, FWS collected information on nonfederal oil and gas activities in the NWRS.[30] Data on nonfederal oil and gas activities presented in this chapter are from the EIS unless otherwise noted.

Within the NWRS, the EIS reported that 107 of the total 605 units,[31] or approximately 18%, contain nonfederally administered wells (including oil, gas, and "other" well types; see *Table 1*).[32]

---

[27] Ibid., p. 79951.

[28] FWS, Service Manual, 612 FW 1, Series: Natural and Cultural Resources Management, Part 612: Mineral Management, December 27, 2016, at https://www.fws.gov/policy/612fw1.pdf

[29] Ibid., Section 1.7(B), at https://www.fws.gov/policy/612fw1.pdf

[30] FWS, "Non-Federal Oil and Gas Development Within the National Wildlife Refuge System," 79 *Federal Register* 10080-10084, February 24, 2014. FWS, *National Wildlife Refuge System Revision of Regulations Governing Non-Federal Oil and Gas Rights: Environmental Impact Analysis*, June 2016.

[31] FWS, *Statistical Data Tables for Fish & Wildlife Service Lands (as of 9/30/2017)*, Tables 3 and 4. The 605 total units include 566 national wildlife refuges, 36 wetland management districts (WMDs), and 3 waterfowl protection areas not otherwise managed by a WMD.

[32] FWS, National Wildlife Refuge System Revision of Regulations Governing Non-Federal Oil and Gas Rights: Environmental Impact Analysis, June 2016, p. 1-6.

Table 1. Number of National Wildlife Refuge System Units with Nonfederal Wells by Region

FWS Region	Units with Oil and/or Gas Wells	Units with "Other" Wells[a]	Total Units with Wells	Total Units in Region[b]
Pacific	0	0	0	68
Southwest	20	16	23	47
Midwest	2	4	5	69
Southeast	20	31	35	130
Northeast	3	3	4	74
Mountain-Prairie	11	14	18	151
Alaska[c]	1	4	4	16
Pacific Southwest	6	18	18	50
Total	63	90	107	605[d]

Sources: Congressional Research Service (CRS), using data from U.S. Fish and Wildlife Service, National Wildlife Refuge System Revision of Regulations Governing Non-Federal Oil and Gas Rights: Environmental Impact Analysis, June 2016, Appendix D, and U.S. Fish and Wildlife Service, Statistical Data Tables for Fish & Wildlife Service Lands (as of 9/30/2017), Tables 3 and 4.

[a] "Other" wells include injection wells for enhanced oil recovery, saltwater or wastewater disposal wells, coalbed methane wells, observation wells, stratigraphic wells, dry wells, and water wells.

[b] Total units include wildlife refuges, wetland management districts, and waterfowl production areas not managed as part of a wetland management district.

[c] See "Oil and Gas Activities in Alaska in the National Wildlife Refuge System" for more information on activities in Alaska.

[d] The 605 total units include 566 national wildlife refuge units, 36 wetland management districts (WMDs), and 3 waterfowl protection areas not otherwise managed by a WMD. Regions were determined by CRS based on the state(s) in which the unit was located. Refuges that spanned region boundaries were counted in a single region.

These units include 104 national wildlife refuges and 3 wetland management districts (*Figure 1*).[33] Forty-five NWRS units were reported to have active wells, including 44 with oil and gas wells and 11 with other well types (some units had both oil and gas wells and other well types).[34] In total, FWS recorded 5,002 nonfederal wells in the NWRS, including 2,201

[33] Ibid., Appendix D.

[34] Pedro Ramirez Jr. and Sherri Baker Mosley, *An Assessment of Oil and Gas Wells and Pipelines on National Wildlife Refuge System Lands*, FWS, June 2014, within FWS, *National Wildlife Refuge System Revision of Regulations Governing Non-Federal Oil and Gas Rights: Environmental Impact Analysis*, June 2016, p. A-10, Table 3.

gas wells, 971 oil wells, 59 oil and gas wells, and 1,771 other wells.[35] "Other" wells include injection wells for enhanced oil recovery, saltwater or wastewater disposal wells, coal-bed methane wells, observation wells, stratigraphic wells, dry wells, and water wells.[36] Of the 5,002 wells, 2,196 (44%) are inactive and 1,665 (33%) are active; the remaining wells have another classification.

For a complete list of NWRS units with nonfederal wells, see *Table A-1*.[37]

Compatibility of Nonfederal Oil and Gas Activities

Balancing mineral owners' right to develop their oil and gas resources with FWS's mission to conserve and protect NWRS resources can be complex and can raise questions of compatible use. The NWRS's mission is provided in the NWRSAA, as amended by the National Wildlife Refuge System Improvement Act of 1997 (P.L. 105-57).[38] The NWRSAA states that "the mission of the System is to administer a national network of lands and waters for the conservation, management, and where appropriate, restoration of the fish, wildlife, and plant resources and their habitats within the United States for the benefit of present and future generations of Americans."[39]

Under the NWRSAA, the Secretary of the Interior is to identify and permit only activities that are deemed compatible with the NWRS's mission and to deny activities that are not compatible.[40]

---

[35] Pedro Ramirez Jr. and Sherri Baker Mosley, *An Assessment of Oil and Gas Wells and Pipelines on National Wildlife Refuge System Lands*, FWS, June 2014, within FWS, *National Wildlife Refuge System Revision of Regulations Governing Non-Federal Oil and Gas Rights: Environmental Impact Analysis*, June 2016, p. A-11, Table 5.

[36] FWS, *National Wildlife Refuge System Revision of Regulations Governing Non-Federal Oil and Gas Rights: Environmental Impact Analysis*, June 2016, p. 1-6.

[37] FWS, *National Wildlife Refuge System Revision of Regulations Governing Non-Federal Oil and Gas Rights: Environmental Impact Analysis*, June 2016, Appendix.

[38] 16 U.S.C. §668dd

[39] 16 U.S.C. §668dd(a)(2).

[40] 16 U.S.C. §668dd(d).

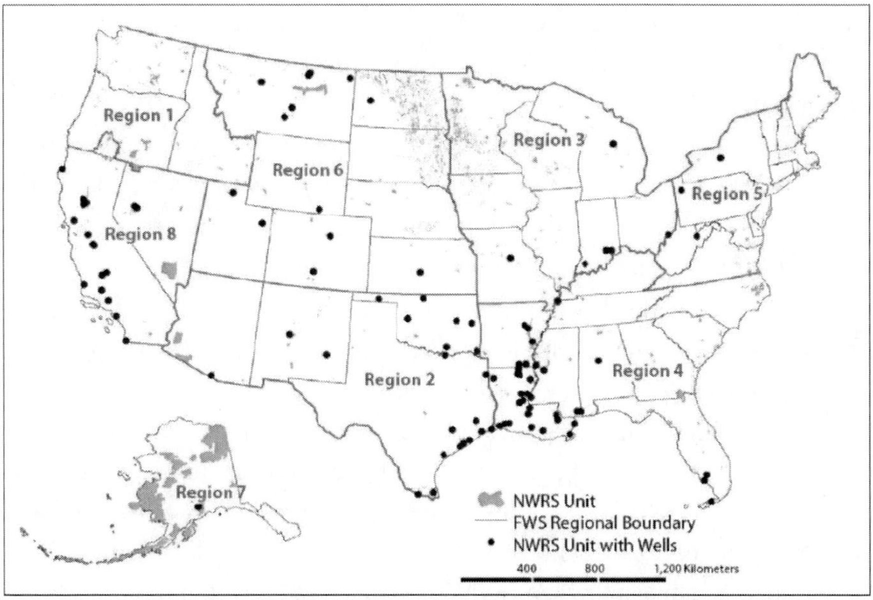

Source: U.S. Fish and Wildlife Service (FWS), National Wildlife Refuge System Revision of Regulations Governing Non-Federal Oil and Gas Rights: Environmental Impact Analysis, June 2016, p. 1-7, Figure 1-1.

Notes: This map includes wildlife refuges and wetland management district units within the National Wildlife Refuge System (NWRS; orange areas). Black dots represent unit locations with both active and/or inactive wells. Whereas the orange areas provide estimated size, the black dots denote location only. FWS's environmental impact statement did not report any wells in FWS units in Hawaii and the U.S. territories, which are not included on the map. Well types include oil, gas, oil and gas, injection wells for enhanced oil recovery, saltwater disposal wells, wastewater disposal wells, coal-bed methane wells, observation wells, stratigraphic wells, dry wells, and water wells used for oil and gas operations. Certain refuges with "other" well types in Table A-1 may not be included this map.

Figure 1. National Wildlife Refuge System Units with Nonfederal Oil and Gas Wells.

Specifically, the law directs the Secretary to facilitate compatible wildlife-dependent recreation opportunities, including hunting, fishing, and hiking. By providing for compatible uses only, the NWRSAA prioritizes the NWRS's primary mission—conservation; management; and restoration of fish, wildlife, and plant resources and their habitats. However, there are limited exceptions to the compatibility requirement; these exceptions provide for certain activities without requiring compatibility determinations.[42]

One such exception is for the development of nonfederal oil and gas rights. FWS's *Final Compatibility Policy Pursuant to the National Wildlife Refuge System Improvement Act of 1997* outlines circumstances in which the compatibility determination process is not required.[43] Exemptions are provided for refuge management activities, military overflights, certain actions taken by other federal agencies with primary jurisdiction over a unit or area within a unit, emergencies, and "other" circumstances, including reserved oil and gas rights. The notice on the policy states, "B. Other exceptions. (1) There are other circumstances under which the compatibility requirements may not be applicable. The most common exceptions involve property rights that are not vested in the Federal Government, such as reserved rights to explore and develop minerals or oil and gas beneath a refuge."[44]

Although these activities are not subject to a compatibility determination, the notice goes on to state, "In the case of reserved rights, the Refuge Manager should work with the owner of the property interest to develop stipulations in a special use permit or other agreement to alleviate or minimize adverse impacts to the refuge."[45]

Questions have been raised about FWS's general authority to regulate nonfederal oil and gas resources within the NWRS.[46] In response to

---

[42] 16 U.S.C. §668dd(d)(4) exempts overflights and certain activities conducted by a non-FWS federal agency that has primary jurisdiction over a unit or section of a unit in the NWRS. §668dd(k) provides an exemption in emergency situations. Additional exemptions are provided for refuge management activities and for activities involving certain reserved nonfederal rights, such as mineral rights, and emergencies, among others. These exemptions are enumerated in Section 2.10 of FWS, "Final Compatibility Policy Pursuant to the National Wildlife Refuge System Improvement Act of 1997," 65 *Federal Register* 62488, October 18, 2000.

[43] FWS, "Final Compatibility Policy Pursuant to the National Wildlife Refuge System Improvement Act of 1997," 65 *Federal Register* 62457-62496, October 18, 2000. These pages contain both the final rule and the associated notice.

[44] Ibid., p. 62488.

[45] Ibid.

[46] FWS, National Wildlife Refuge System Revision of Regulations Governing Non-Federal Oil and Gas Activities, Advance Notice of Proposed Rulemaking—Public Comment Analysis,

questions raised in public comments during development of the 2016 rule, FWS cited its statutory authority to regulate nonfederal oil and gas resources within the NWRS through the NWRSAA.[47] FWS relies on this same authority in requiring permits for oil and gas operations within the NWRS. The agency states that the development of oil and gas resources within the NWRS has the potential to "adversely impact [other] refuge resources."[48]

FEDERAL OIL AND GAS OPERATIONS IN THE NATIONAL WILDLIFE REFUGE SYSTEM

Under the Mineral Leasing Act of 1920, BLM manages oil and gas leasing on federal lands, including in the NWRS. BLM regulations provide that any drilling in national wildlife refuges may take place only "with the consent and approval of the Secretary [of the Interior] with the concurrence of the Fish and Wildlife Service as to the time, place and nature of such operations in order to give complete protection to wildlife populations and wildlife habitat on the areas leased."[49]

FWS policy concerning federal oil and gas leasing within the NWRS is summarized in the agency's *Mineral Policy Service Manual*.[50] The manual states that, with limited exceptions, extracting or developing leasable federal minerals is not allowed on NWRS lands. The exception to this prohibition is for limited circumstances where oil and gas development adjacent to an NWRS unit results in the drainage of federal oil and gas

December 1, 2014, at https://www.fws.gov/refuges/oil-and-gas/pdfs/ANPR-Public-Comment-Report-Analysis.pdf.

[47] 16 U.S.C. §668dd(b)(5) and §668dd(d)(1)(a).

[48] FWS, "Management of Non-Federal Oil and Gas Rights," 81 Federal Register 79949-79950, November 14, 2016.

[49] 43 C.F.R. §3101.5.

[50] FWS, *Service Manual*, 612 FW 1, Series: Natural and Cultural Resources Management, Part 612: Mineral Management, December 27, 2016, Section 1.7(A), at https://www.fws.gov/policy/612fw1.pdf.

resources.[51] In cases where the federal leases predate the establishment or expansion of the refuge, valid existing leases and wells may continue to operate.[52]

Number and Location of Federal Operations

BLM reports that it administers 328 federal wells in 11 NWRS units outside of Alaska and 1 unit in Alaska. (The total of 12 units includes 10 national wildlife refuges and 2 wetland management districts. See "Oil and Gas Activities in Alaska in the National Wildlife Refuge System" for more information on activities in Alaska.)[53] Of the 11 units outside of Alaska, the majority also have nonfederal oil and gas wells. However, two national wildlife refuges and one wetland management district that contain federal wells do not have nonfederal wells. The total includes 35 producing oil wells, 65 producing gas wells, and 194 nonproducing wells of various types.[54] In addition, 34 wells are in stages of the approval process within these units.[55] The wells are located in seven states within the southeast (Louisiana), mountain (Montana, Utah, and North Dakota), southwest (New Mexico and Oklahoma), and midwest (Indiana) regions. See *Table B-1* for a complete list of federal oil and gas activity within the NWRS.

---

[51] FWS *Service Manual*, Part 612, Section 1.7(A)(2), states, "Leasing of Federal minerals is generally prohibited. Leasing of Federal oil and gas is only allowed if there is drainage of Federal oil and gas by adjacent development. The Bureau of Land Management (BLM) conducts the leasing in this situation, but it can only occur with the approval of the Secretary of the Interior, and with the Service's concurrence as to the time, place, and nature of those operations (see 43 C.F.R. 3101.5-1)."

[52] FWS, NWRS, "Oil and Gas."

[53] Personal communication between Marc Humphries, CRS Specialist in Energy Policy, and Bureau of Land Management Office of Legislative and Congressional Affairs, February 9, 2018.

[54] Non-producing wells include abandoned, gas shut-in, oil shut-in, plugged and abandoned, gas injection, temporarily abandoned, water disposal, and water injection wells.

[55] The stages of the approval process are approved (1 well), unapproved (32 wells), and pending notice of staking (1 well).

Compatibility of Federal Oil and Gas Activities

Because new development of federal oil and gas resources on refuge lands typically is prohibited, in undertaking such evaluations, it is unclear whether the FWS "concurrence" required in BLM's oil and gas leasing regulations would involve an FWS compatibility determination.[56] As discussed above, FWS's compatibility policy describes some situations in which a compatibility determination is not required.[57] Among other situations, exceptions may apply "where primary jurisdiction of refuge lands falls to an agency other than [FWS]." It is unclear whether this exception would apply in the case of federal subsurface resources, which are administered by BLM.[58]

OIL AND GAS ACTIVITIES IN ALASKA IN THE NATIONAL WILDLIFE REFUGE SYSTEM

National wildlife refuges in Alaska are governed by the provisions of ANILCA and other Alaska-specific laws, as well as by general authorities for the NWRS.[59] ANILCA provides FWS with statutory authority to ensure that operators accessing nonfederal mineral rights within NWRS in Alaska work with FWS to help avoid operational impacts and ensure reclamation standards.[60] As noted, the regulations promulgated in the FWS nonfederal

---

[56] CRS communication with FWS, November 8, 2017.

[57] FWS, "Final Compatibility Policy Pursuant to the National Wildlife Refuge System Improvement Act of 1997," 65 *Federal Register* 62457-62496, October 18, 2000. Note: citation includes page range for both the final rule and the associated notice; both contain the same title.

[58] For Alaska, state-specific statutes and regulations create a different framework. See "Compatibility of Oil and Gas Activities in Alaska."

[59] Specifically, Title XI of the Alaska National Interest Lands Conservation Act (ANILCA; P.L. 96-497, 16 U.S.C. §§3161-3173) and §22(g) of the Alaska Native Claims Settlement Act (ANCSA; 43 U.S.C. §1621(g)), as well as associated implementing regulations, govern the administration of NWRS lands in Alaska. Also see FWS, *Service Manual, 612 FW 1, Series: Natural and Cultural Resources Management, Part 612: Mineral Management*, December 27, 2016, Section 1.7(F), at https://www.fws.gov/policy/612fw1.pdf

[60] FWS *Service Manual*, Part 612, Section 1.7(F)(2) states: "Project Leaders should use the performance-based standards in 50 CFR Part 29 Subpart D, as appropriate, to help them

oil and gas rule of November 2016 do *not* apply to refuges in Alaska.[61] Although Alaska is exempted from regulations within the rule, the rule states

> Operations on refuges in Alaska are exempt from these regulations. However, the performance-based standards of this rule may be used, as appropriate, as guidance in determining how an operator would meet the various requirements of ANILCA and ANCSA [Alaska Native Claims Settlement Act, 43 U.S.C. §§1601 et seq.] to protect refuge resources and uses.[62]

Number and Location of Oil and Gas Operations in Alaska Refuges

With respect to *nonfederal* oil and gas operations in Alaska refuges, FWS identified four NWRS units in Alaska with nonfederal wells (see *Table 1* and *Table A-1*). Nonfederal subsurface ownership in Alaska may arise from situations in which FWS acquired split-estate lands without subsurface rights or from situations in which subsurface rights were legislatively conveyed under ANCSA.[63]

In total, FWS identified 199 nonfederal Alaska wells, including gas, oil, oil and gas, and other well types. Of the 199 wells, 191 are contained within a single unit, Kenai National Wildlife Refuge, and the remaining 8 (all noted as "other" for their type) are distributed across the remaining three refuges. Only Kenai National Wildlife Refuge contains both active and inactive wells; all wells in Alaska Peninsula, Becharof, and Yukon Delta National Wildlife Refuges are inactive.[64]

---

determine how an operator would meet the various requirements of ANILCA and ANCSA to protect NWRS resources and uses for non-Federal oil and gas."

[61] FWS, "Management of Non-Federal Oil and Gas Rights," 81 *Federal Register* 79951, November 14, 2016.

[62] Ibid.

[63] FWS, *Non-Federal Oil & Gas Activities on National Wildlife Refuge System Lands*, November 2016, at https://www.fws.gov/refuges/oil-and-gas/pdfs/Oil-Gas-Fact-sheet.pdf.

[64] FWS, *National Wildlife Refuge System Revision of Regulations Governing Non-Federal Oil and Gas Rights: Environmental Impact Analysis*, June 2016, p. A-10, Table 3.

With respect to *federal* oil and gas operations in Alaska refuges, BLM identified 92 wells, all within the Kenai National Wildlife Refuge.[65] Of this total, 30 wells are producing oil, 12 wells are producing gas, and 49 are nonproducing wells of various types. One additional well is in the drilling stage. As discussed above (see "Arctic National Wildlife Refuge"), P.L. 115-97 requires the development of a federal oil and gas program within the ANWR.

See *Table B-1* for a complete list of federal wells in Alaska.

Compatibility of Oil and Gas Activities in Alaska

Section 304 of ANILCA states: "Each refuge shall be administered by the Secretary [of the Interior], subject to valid existing rights, in accordance with the laws governing the administration of units of the National Wildlife Refuge System, and this Act."[66] The section additionally states that the Secretary must determine whether proposed refuge uses (including certain oil and gas activities discussed in ANILCA) are compatible with refuge purposes before permitting those uses.[67] FWS states in its *Service Manual* that the agency "may alter the compatibility process, in some cases, for Alaska refuges," including for reviewing oil and gas leasing on non-North Slope refuges.[68] Also, BLM regulations at 43 C.F.R. 3101.5-3 state with regard to Alaska wildlife areas that "no lands within a refuge in Alaska open to leasing shall be available until the Fish and Wildlife Service has first completed compatability [*sic*] determinations."[69]

Section 1008 of ANILCA outlines a requirement for the Secretary of the Interior to establish an oil and gas leasing program for certain non-

[65] Personal communication between Marc Humphries, CRS Specialist in Energy Policy, and Bureau of Land Management Office of Legislative and Congressional Affairs, February 9, 2018.

[66] ANILCA §304.

[67] ANILCA §304(b).

[68] FWS, Service Manual, 603 FW 2, Series: Refuge Management, Part 603: National Wildlife Refuge System Uses, November 17, 2000, Section 2.6(A), at https://www.fws.gov/policy/603fw2.pdf. The North Slope consists of land in Alaska that lies north of the crest of the Brooks Mountain Range.

[69] 43 C.F.R. §3101.5-3.

North Slope federal lands in Alaska.[70] The section provides that the Secretary shall not identify lands for oil and gas development where this activity would be in violation of applicable law or, in the case of NWRS lands, would be incompatible with the purpose for which the units were established. Section 1009 requires that, upon receipt of an application for an oil and gas lease, the Secretary must render a decision, accompanied by a statement explaining the decision and whether oil and gas activities would be compatible with the purpose of the NWRS unit, within six months or within three months after publication of a final environmental impact statement if the Secretary determines one is required under the National Environmental Policy Act (NEPA; 42 U.S.C. §§4321-4347).[71]

In debating oil and gas development in the ANWR, Congress has considered the compatibility of oil and gas activities with refuge purposes. P.L. 115-97, which established an oil and gas program for ANWR's Coastal Plain, amended ANILCA to include a new refuge purpose: "to provide for an oil and gas program on the Coastal Plain." Including oil and gas as a refuge purpose appears likely to facilitate a compatibility decision for an oil and gas program. FWS would prepare a compatibility determination as part of BLM's development of the oil and gas program, which has not yet occurred.

POTENTIAL ISSUES FOR CONGRESS

The 115th Congress has considered several issues relating to oil and gas activities in the NWRS. Issues of interest have included the general compatibility of oil and gas activity with NWRS purposes and oil and gas development in ANWR specifically. (See "Recent Developments" section for more information.)

H.J.Res. 45 was introduced, under the authorities provided by the Congressional Review Act (CRA), to disapprove the 2016 final rule pertaining to nonfederal oil and gas activities within the NWRS outside of

[70] ANILCA §1008. Also at 16 U.S.C. §3148.
[71] ANILCA §1009. Also at 16 U.S.C. §3149.

Alaska.[72] Although the window for considering H.J.Res. 45 under the expedited procedures associated with the CRA has elapsed, Congress may consider the issues raised in this legislation. For example, Congress may be interested in issues pertaining to permitting fees, the role of state regulators, and how to balance or prioritize facilitation of activities with concerns about wildlife and resource protection.

Separately in the 115th Congress, P.L. 115-97 addressed the long-standing issue of energy development in ANWR by establishing an oil and gas program for the Coastal Plain within the refuge.[73] H.R. 49 and S. 49, introduced prior to the passage of P.L. 115-97, proposed oil and gas leasing programs for the Coastal Plain that contained similar but not identical provisions to the enacted law. Also introduced in the 115th Congress, H.R. 1889 and S. 820 would designate the Coastal Plain as wilderness under the Wilderness Act, which generally would prohibit commercial activities, including energy development.[74]

Congress may remain interested in key aspects of the ANWR debate that have been raised in P.L. 115-97 and previously. These could include issues related to limits on the footprint of development, other environmental protections, compliance with NEPA, FWS and BLM management roles, judicial review of legal challenges, and treatment of special areas within the Coastal Plain, among other matters. P.L. 115-97 addressed some of these issues and not others, possibly owing in part to limitations imposed by the budget reconciliation process on the matters that can be considered germane.[75] Congress could choose to address these issues through future oversight or legislation.

---

[72] FWS, "Management of Non-Federal Oil and Gas Rights," 81 *Federal Register* 79948-79981, November 14, 2016. This rule became effective on December 14, 2016, updating a previous rule.

[73] For more information on the Arctic National Wildlife Refuge, see CRS Report RL33872, *Arctic National Wildlife Refuge (ANWR): An Overview*, by Laura B. Comay, Michael Ratner, and R. Eliot Crafton.

[74] 16 U.S.C. §§1131 et seq., Wilderness Act. For more information, see CRS Report RL31447, *Wilderness: Overview, Management, and Statistics*, by Katie Hoover.

[75] In particular, the Senate's "Byrd rule" limits the inclusion of provisions extraneous to achieving the goals of the reconciliation instructions, including provisions that are outside the jurisdiction of the committee submitting the reconciliation measure. For more information, see CRS Report RL30862, *The Budget Reconciliation Process: The Senate's "Byrd Rule,"* by Bill Heniff Jr.

APPENDIX A. NONFEDERAL OIL AND GAS ACTIVITIES WITHIN THE NATIONAL WILDLIFE REFUGE SYSTEM

Table A-1. National Wildlife Refuge System Units with Nonfederal Oil and Gas Wells **(number of wells)**

Region / State / NWRS Unit	Gas	Oil	Oil and Gas	Other[a]	Total
Region 1 – Pacific	0	0	0	0	0
Region 2 – Southwest	358	282	26	308	974
Arizona	—	—	—	1	1
Buenos Aires National Wildlife Refuge	—	—	—	1	1
New Mexico	5	10	—	2	17
Bitter Lake National Wildlife Refuge	5	10	—	—	15
Sevilleta National Wildlife Refuge	—	—	—	2	2
Oklahoma	62	149	—	255	466
Deep Fork National Wildlife Refuge	26	147	—	227	400
Little River National Wildlife Refuge	—	1	—	7	8
Optima National Wildlife Refuge	13	—	—	2	15
Salt Plains National Wildlife Refuge	11	1	—	1	13
Sequoyah National Wildlife Refuge	—	—	—	4	4
Tishomingo National Wildlife Refuge	2	—	—	12	14
Washita National Wildlife Refuge	10	—	—	2	12
Texas	291	123	26	50	490
Anahuac National Wildlife Refuge	5	7	—	3	15
Aransas National Wildlife Refuge	58	8	4	2	72
Attwater Prairie Chicken National Wildlife Refuge	19	1	—	—	20
Big Boggy National Wildlife Refuge	1	—	—	—	1
Brazoria National Wildlife Refuge	12	4	3	—	19
Caddo Lake National Wildlife Refuge	2	2	—	1	5
Hagerman National Wildlife Refuge	3	71	4	38	116
Laguna Atascosa National Wildlife Refuge	11	—	—	—	11
Lower Rio Grande Valley National Wildlife Refuge	139	23	10	3	175
McFaddin National Wildlife Refuge	14	4	—	2	20
San Bernard National Wildlife Refuge	24	3	2	1	30
Texas Point National Wildlife Refuge	—	—	2	—	2
Trinity River National Wildlife Refuge	3	—	1	—	4
Region 3 – Midwest	2	7	—	93	102
Indiana	—	7	—	89	96
Big Oaks National Wildlife Refuge Muscatatuck	—	—	—	5	5
National Wildlife Refuge Patoka River National	—	—	—	1	1

Region / State / NWRS Unit	**Gas**	**Oil**	**Oil and Gas**	**Other[a]**	**Total**
Wildlife Refuge	—	7	—	83	90
Michigan	2	—	—	—	2
Kirtlands Warbler Wildlife Management Area	2	—	—	—	2
Missouri	—	—	—	4	4
Big Muddy National Fish And Wildlife Refuge	—	—	—	4	4
Region 4 – Southeast	1,709	530	16	1,172	3,427
Alabama	—	—	—	14	14
Cahaba River National Wildlife Refuge	—	—	—	14	14
Arkansas	8	51	—	6	65
Bald Knob National Wildlife Refuge	3	—	—	—	3
Cache River National Wildlife Refuge	2	—	—	—	2
Felsenthal National Wildlife Refuge	3	48	—	6	57
Overflow National Wildlife Refuge	—	1	—	—	1
White River National Wildlife Refuge	—	2	—	—	2
Florida	—	—	—	4	4
Florida Panther National Wildlife Refuge	—	—	—	2	2
National Key Deer Refuge	—	—	—	1	1
Ten Thousand Islands National Wildlife Refuge	—	—	—	1	1
Kentucky	—	—	—	1	1
Reelfoot National Wildlife Refuge	—	—	—	1	1
Louisiana	1,693	349	16	772	2,830
Atchafalaya National Wildlife Refuge	9	12	2	23	46
Bayou Cocodrie National Wildlife Refuge	—	—	—	49	49
Bayou Sauvage National Wildlife Refuge	—	—	—	4	4
Bayou Teche National Wildlife Refuge	14	3	—	18	35
Big Branch Marsh National Wildlife Refuge	—	—	—	4	4
Black Bayou Lake National Wildlife Refuge	78	—	—	7	85
Breton National Wildlife Refuge	—	—	—	3	3
Cameron Prairie National Wildlife Refuge	—	—	—	14	14
Cat Island National Wildlife Refuge	2	—	—	3	5
Catahoula National Wildlife Refuge	1	20	—	54	75
D'Arbonne National Wildlife Refuge	183	—	—	100	283
Delta National Wildlife Refuge	52	214	11	86	363
Grand Cote National Wildlife Refuge	—	—	—	2	2
Lacassine National Wildlife Refuge	32	5	1	42	80
Lake Ophelia National Wildlife Refuge	—	5	—	51	56
Mandalay National Wildlife Refuge	13	5	—	25	43
Red River National Wildlife Refuge	6	24	—	27	57
Sabine National Wildlife Refuge	22	23	2	51	98

Table A-1. (Continued)

Region / State / NWRS Unit	Gas	Oil	Oil and Gas	Other[a]	Total
St. Catherine Creek National Wildlife Refuge	—	—	—	6	6
Tensas River National Wildlife Refuge	1	36	—	75	112
Upper Ouachita National Wildlife Refuge	1,280	2	—	128	1,410
Mississippi	8	130	—	375	513
Grand Bay National Wildlife Refuge	—	—	—	1	1
Mississippi Sandhill Crane National Wildlife Refuge	—	—	—	1	1
Panther Swamp National Wildlife Refuge	—	—	—	2	2
St. Catherine Creek National Wildlife Refuge	8	130	—	370	508
Yazoo National Wildlife Refuge	—	—	—	1	1
Region 5 – Northeast	7	4	11	28	50
New York	—	—	—	1	1
Montezuma National Wildlife Refuge	—	—	—	1	1
Pennsylvania	2	—	—	—	2
Erie National Wildlife Refuge	2	—	—	—	2
West Virginia	5	4	11	27	47
Canaan Valley National Wildlife Refuge	4	—	—	3	7
Ohio River Islands National Wildlife Refuge	1	4	11	24	40
Region 6 – Mountain-Prairie	42	37	2	41	122
Colorado	—	—	—	3	3
Baca National Wildlife Refuge	—	—	—	2	2
Rocky Mountain Arsenal National Wildlife Refuge	—	—	—	1	1
Kansas	1	25	1	—	27
Quivira National Wildlife Refuge	1	25	1	—	27
Montana	40	9	—	34	83
Benton Lake National Wildlife Refuge	—	—	—	2	2
Benton Lake Wetland Management District	4	—	—	7	11
Bowdoin National Wildlife Refuge	1	—	—	1	2
Bowdoin Wetland Management District	21	—	—	2	23
Hailstone National Wildlife Refuge	—	—	—	1	1
Halfbreed Lake National Wildlife Refuge	1	—	—	4	5
Hewitt Lake National Wildlife Refuge	13	—	—	1	14
Lake Mason National Wildlife Refuge	—	—	—	2	2
Medicine Lake National Wildlife Refuge	—	2	—	3	5
Northeast Montana Wetland Management District	—	7	—	11	18
North Dakota	—	—	1	—	1
Lake Ilo National Wildlife Refuge	—	—	1	—	1

Region / State / NWRS Unit	**Gas**	**Oil**	**Oil and Gas**	**Other[a]**	**Total**
Utah	1	3	—	3	7
Bear River Migratory Bird Refuge	—	—	—	3	3
Colorado River Wildlife Management Area	1	2	—	—	3
Ouray National Wildlife Refuge	—	1	—	—	1
Wyoming	—	—	—	1	1
Mortenson Lake National Wildlife Refuge	—	—	—	1	1
Region 7 – Alaska	75	66	4	54	199
Alaska	75	66	4	54	199
Alaska Peninsula National Wildlife Refuge	—	—	—	5	5
Becharof National Wildlife Refuge	—	—	—	2	2
Kenai National Wildlife Refuge	75	66	4	46	191
Yukon Delta National Wildlife Refuge	—	—	—	1	1
Region 8 – Pacific Southwest	8	45	—	75	128
California	8	45	—	70	123
Bitter Creek National Wildlife Refuge	—	—	—	12	12
Butte Sink Wildlife Management Area	—	—	—	1	1
Colusa National Wildlife Refuge	1	—	—	3	4
Delevan National Wildlife Refuge	1	—	—	7	8
Grasslands Wildlife Management Area	—	—	—	6	6
Guadalupe-Nipomo Dunes National Wildlife Refuge	—	2	—	1	3
Hopper Mountain National Wildlife Refuge	—	17	—	2	19
Humboldt Bay National Wildlife Refuge	—	—	—	1	1
Kern National Wildlife Refuge	—	—	—	2	2
Merced National Wildlife Refuge	—	—	—	1	1
North Central Valley Wildlife Management Area	6	—	—	22	28
Pixley National Wildlife Refuge	—	—	—	1	1
San Joaquin River National Wildlife Refuge	—	—	—	5	5
San Pablo Bay National Wildlife Refuge	—	—	—	1	1
Seal Beach National Wildlife Refuge	—	26	—	4	30
Tijuana Slough National Wildlife Refuge	—	—	—	1	1
Nevada	—	—	—	5	5
Fallon National Wildlife Refuge	—	—	—	1	1
Stillwater National Wildlife Refuge	—	—	—	4	4
Grand Total	2,201	971	59	1,771	5,002

Source: CRS, with data from U.S. Fish and Wildlife Service (FWS), *National Wildlife Refuge System Revision of Regulations Governing Non-Federal Oil and Gas Rights: Environmental Impact Analysis*, June 2016, Appendix D.

Notes: The table represents the best available data as analyzed for FWS's environmental impact statement.

[a] The "Other" category includes wells other than oil and gas: injection, water disposal, enhanced oil recovery, dry, observation, stratigraphic, other, and production type data not available.

Appendix B. Federal Oil and Gas Wells Within the National Wildlife Refuge System

Table B-1. National Wildlife Refuge System Units with Federal Oil and Gas Wells (number of wells)

Region / State / NWRS Unit	**Not Producing**[a]		**Producing**			**Not Drilled**[b]		**Total**
Gas	Oil		Other	Gas		Oil		Drilling
Region 1 –Pacific	0	0	0	0	0	0	0	0
Region 2 –Southwest	0	0	7	29	6	0	14	56
New Mexico	—	—	7	*27*	*1*	—	13	48
Bitter Lake National Wildlife Refuge	—	—	7	27	1	—	13	48
Oklahoma	—	—	—	*2*	*5*	—	1	8
Salt Plains National Wildlife Refuge	—	—	—	2	5	—	1	8
Region 3 –Midwest	0	0	5	0	2	0	0	7
Indiana	—	—	5	—	*2*	—	—	7
Patoka River National Wildlife Refuge	—	—	5	—	2	—	—	7
Region 4 –Southeast	1	2	154	7	16	0	6	186
Louisiana	*1*	*2*	154	*7*	*16*	—	6	186
Delta-Breton National Wildlife Refuge	1	2	154	7	15	—	6	185
Tern Island National Wildlife Refuge	—	—	0	—	1	—	0	1
Region 5 –Northeast	0	0	0	0	0	0	0	0
Region 6 –Mountain-Prairie	1	0	24	29	11	0	14	79
Montana	1	—	10	27	1	—	1	40
Bowdoin National Wildlife Refuge	1	—	8	23	—	—	1	33
Hewitt Lake National Wildlife Refuge	—	—	2	4	—	—	—	6
Northeast Montana Wetland Management District (Sheridan County Waterfowl Production Area)	—	—	—	—	1	—	—	1
North Dakota	—	—	—	—	7	—	—	7
Upper Souris National Wildlife Refuge	—	—	—	—	6	—	—	6
Crosby Wetland Management District (Williams County Waterfowl Production Area)	—	—	—	—	1	—	—	1

Region / State / NWRS Unit	**Not Producing**[a]		**Producing**			**Not Drilled**[b]		**Total**
Gas	Oil		Other	Gas		Oil		Drilling
Utah	—	—	14	2	3	—	13	32
Ouray National Wildlife Refuge	—	—	14	2	3	—	13	32
Region 7 –Alaska	7	4	38	12	30	1	0	92
Alaska	7	4	38	12	30	1	—	92
Kenai National Wildlife Refuge	7	4	38	12	30	1	—	92
Region 8 –Pacific Southwest	0	0	0	0	0	0	0	0
Grand Total	9	6	228	77	65	1	34	420

Source: CRS, with data from personal communication between Marc Humphries, Specialist in Energy Policy, and Bureau of Land Management Office of Legislative and Congressional Affairs, February 9, 2018.

[a] Not-producing well types: gas (gas shut-in); oil (oil shut-in); other: abandoned, plugged and abandoned, gas injection, temporarily abandoned, water disposal, and water injection. A shut-in well is one that has produced oil or gas and is capable of producing but is not currently doing so.

[b] Not drilled includes proposed wells with statuses of approved, unapproved, and pending notice of staking.

In: Energy Policies: Issues and Legislation ISBN: 978-1-53614-485-7
Editor: Doug Higgins

Chapter 3

NUCLEAR COOPERATION WITH OTHER COUNTRIES: A PRIMER (UPDATED)*

Paul K. Kerr and Mary Beth D. Nikitin

SUMMARY

In order for the United States to engage in significant civilian nuclear cooperation with other states, it must conclude a framework agreement that meets specific requirements under Section 123 of the Atomic Energy Act (AEA). Significant nuclear cooperation includes the export of reactors, critical parts of reactors, and reactor fuel. The AEA also provides for export control licensing procedures and criteria for terminating cooperation. Congressional review is required for Section 123 agreements; the AEA establishes special parliamentary procedures by which Congress may act on a proposed agreement.

* This is an edited, reformatted and augmented version of a Congressional Research Service report, RS22937, dated April 3, 2018.

WHAT IS A "SECTION 123" AGREEMENT?

Under existing law (Atomic Energy Act [AEA] of 1954, as amended; (P.L. 83-703; 42 U.S.C. §2153 et seq.)[1] all significant U.S. nuclear cooperation with other countries requires a peaceful nuclear cooperation agreement.[2] Significant nuclear cooperation includes the transfer of U.S.-origin special nuclear material[3] subject to licensing for commercial, medical, and industrial purposes, and the export of reactors and critical parts of reactors. Section 123 agreements are required for the export of commodities under NRC export licensing authority (10 C.F.R. 110).[4]

Such agreements, which are "congressional-executive agreements" requiring congressional approval, do not guarantee that cooperation will take place or that nuclear material will be transferred, but rather set the terms of reference and authorize cooperation. The AEA includes requirements for an agreement's content, conditions for the President to

---

[1] The Atomic Energy Act (AEA) was amended by the Nuclear Nonproliferation Act of 1978 (NNPA) (P.L. 95-242) to include stringent nonproliferation requirements for significant U.S. nuclear exports. For example, the act required nonnuclear weapon states to have full-scope International Atomic Energy Agency safeguards as a condition for entering into nuclear cooperation agreements with the United States. For existing and future agreements, the NNPA added a provision for Congress to review export licenses. The act also included a provision for halting exports if a country tested a nuclear device, violated safeguards agreements, or continued nuclear weapons-related activities.

[2] Section 57b. (2) of the AEA allows for limited forms of nuclear cooperation related to the "development or production of any special nuclear material outside of the United States" without a nuclear cooperation agreement if that activity has been authorized by the Secretary of Energy following a determination that it "will not be inimical to the interest of the United States." Agreements governing such cooperation are also known as "Section 810" agreements, after 10 Code of Federal Regulations Part 810. (See "Part 810 Agreements.")

A nuclear cooperation agreement is not required for transmission of nuclear-related information, except for restricted data. "Restricted data," defined by the statute, means "all data concerning (1) design, manufacture, or utilization of atomic weapons; (2) the production of special nuclear material; or (3) the use of special nuclear material in the production of energy." Restricted data, however does not "include data declassified or removed from the Restricted Data [sic] category" pursuant to the AEA. A nuclear cooperation agreement is necessary, though not necessarily sufficient, to permit the transfer of restricted data.

[3] "Special nuclear material," defined by the statute, means (1) plutonium, uranium enriched in the isotopes 233 or 235, and any other material that is determined to be special nuclear material, but does not include source material, or (2) any material artificially enriched by any of the foregoing, but does not include source material.

[4] For a list of commodities, see http://www.nrc.gov/about-nrc/ip/export-import.html.

exempt an agreement from those requirements, presidential determinations and other supporting information to be submitted to Congress, conditions affecting the implementation of an agreement once it takes effect, and procedures for Congress to consider and approve the agreement.

Section 123 of the AEA requires that any agreement for nuclear cooperation meet nine nonproliferation criteria and that the President submit any such agreement to the House Committee on Foreign Affairs and the Senate Committee on Foreign Relations. The Department of State is required to provide the President with an unclassified Nuclear Proliferation Assessment Statement (NPAS), which the President is to submit, along with the agreement, to those two committees. The State Department is also required to provide a classified annex to the NPAS, prepared in consultation with the Director of National Intelligence. The NPAS is meant to explain how the agreement meets the AEA nonproliferation requirements. The President must also make a written determination "that the performance of the proposed agreement will promote and will not constitute an unreasonable risk to, the common defense and security."

REQUIREMENTS UNDER THE ATOMIC ENERGY ACT

Section 123 of the AEA specifies the necessary steps for engaging in nuclear cooperation with another country.

- *Section 123a.* states that the proposed agreement is to include the terms, conditions, duration, nature, and scope of cooperation and lists nine criteria that the agreement must meet. It also contains provisions for the President to exempt an agreement from any of several criteria described in that section and includes details on the kinds of information the executive branch must provide to Congress.
- *Section 123b.* specifies the process for submitting the text of the agreement to Congress.

- *Section 123c.* specifies the procedure for congressional approval of cooperation agreements that are limited in scope (e.g., do not transfer nuclear material or cover reactors larger than 5 megawatts electric [MWe]). This chapter does not discuss such agreements.
- *Section 123d.* specifies the procedure for congressional approval of agreements that do cover significant nuclear cooperation (transfer of nuclear material or reactors larger than 5 MWe), including exempted agreements.

Section 123a., paragraphs (1) through (9), lists nine criteria that an agreement with a non-nuclear weapon state must meet unless the President determines an exemption is necessary. These include guarantees that

- safeguards on transferred nuclear material and equipment continue in perpetuity;
- International Atomic Energy Agency (IAEA) comprehensive safeguards are applied in non-nuclear weapon states;
- nothing transferred is used for any nuclear explosive device or for any other military purpose; the United States has the right to demand the return of transferred nuclear materials and equipment, as well as any special nuclear material produced through their use, if the cooperating state detonates a nuclear explosive device or terminates or abrogates an IAEA safeguards agreement;
- there is no retransfer of material or classified data without U.S. consent;
- physical security on nuclear material is maintained;
- there is no enrichment or reprocessing by the recipient state of transferred nuclear material or nuclear material produced with materials or facilities transferred pursuant to the agreement without prior approval;
- storage for transferred plutonium and highly enriched uranium is approved in advance by the United States; and

- any material or facility produced or constructed through use of special nuclear technology transferred under the cooperation agreement is subject to all of the above requirements.

Although some experts have advocated requiring governments to forgo enrichment and reprocessing (a nonproliferation commitment sometimes referred to as the "Gold Standard") as a condition for concluding a nuclear cooperation agreement, the Atomic Energy Act does include such a requirement (see *Appendix B*).

Exempted vs. Non-Exempted Agreements

The President may exempt an agreement for cooperation from any of the requirements in Section 123a. if he determines that the requirement would be "seriously prejudicial to the achievement of U.S. non-proliferation objectives or otherwise jeopardize the common defense and security." The AEA provides different requirements, conditions, and procedures for exempt and non-exempt agreements.[5] To date, all of the Section 123 agreements in force are non-exempt agreements.[6] Prior to the adoption of P.L. 109-401, the Henry J. Hyde United States-India Peaceful Atomic Energy Cooperation Act of 2006, the President would have needed to exempt the nuclear cooperation agreement with India from some requirements of Section 123a. However, P.L. 109- 401 exempted nuclear cooperation with India from some of the AEA's requirements.[7]

---

[5] Nuclear cooperation agreements with nuclear weapon states recognized by the NPT are provided for in the AEA, and are therefore non-exempt agreements. The NPT defines nuclear weapon states as those that exploded a nuclear weapon or other nuclear explosive device prior to January 1, 1967: China, France, Russia, the United Kingdom, and the United States.

[6] The United States has concluded more than 20 bilateral nuclear cooperation agreements, as well as similar agreements with the European Atomic Energy Community and the IAEA. See *Nuclear Commerce: Government-wide Strategy Could Help Increase Commercial Benefits from U.S. Nuclear Cooperation Agreements with Other Countries*, Government Accountability Office, GAO-11-36, November 2010.

[7] See CRS Report RL33016, *U.S. Nuclear Cooperation with India: Issues for Congress*, by Paul K. Kerr. The agreement entered into force in December 2008.

Congressional Review

Under the AEA, Congress has the opportunity to review a nuclear cooperation agreement for two time periods totaling 90 days of continuous session.[8] The President must submit the text of the proposed agreement, along with required supporting documents (including the unclassified NPAS) to the House Foreign Affairs Committee and the Senate Foreign Relations Committee. The President is to consult with the committees "for a period of not less than 30 days of continuous session." After this period of consultation, the President is to submit the agreement to Congress, along with the classified annex to the NPAS and a statement of his approval of the agreement and determination that it will not damage U.S. national security interests. This action begins the second period, which consists of 60 days of continuous session. In practice, the President has sent the agreement to Congress at the beginning of the full 90-day period, which begins on the date of transmittal. Typically, the 60-day period has immediately followed the expiration of the 30-day period. The President transmits the text of the proposed agreement along with a letter of support with a national security determination, the unclassified NPAS, its classified annex, and letters of support for the agreement from the Secretary of State and the Nuclear Regulatory Commission.

If the President has not exempted the agreement from any requirements of Section 123a., it may enter into force after the end of the 60-day period unless, during that time, Congress adopts a joint resolution disapproving the agreement and the resolution becomes law. If the agreement is an exempted agreement, Congress must adopt a joint resolution of approval and it must become law by the end of the 60-day period or the agreement may not enter into force. At the beginning of this 60-day period, joint resolutions of approval or disapproval, as appropriate,

---

[8] When calculating periods of "continuous session" under the AEA, every calendar day is counted, including Saturdays and Sundays. Only days on which either chamber has adjourned for more than three days pursuant to the adoption a concurrent resolution authorizing the adjournment do not count toward the total. If Congress adjourns its final session *sine die*, continuity of session is broken, and the count must start anew when it reconvenes.

are to be automatically introduced in each house. During this period, the committees are to hold hearings on the proposed agreement and "submit a report to their respective bodies recommending whether it should be approved or disapproved." If either committee has not reported the requisite joint resolution of approval or disapproval by the end of 45 days, it is automatically discharged from further consideration of the measure. After the joint resolution is reported or discharged, Congress is to consider it under expedited procedures, as established by Section 130.i. of the AEA. Congress has used procedures outside the above-described process to adopt legislation approving some nuclear cooperation agreements (see *Appendix C*).

Section 202 of P.L. 110-369, the United States-India Nuclear Cooperation Approval and Nonproliferation Enhancement Act, which President Bush signed into law October 8, 2008, amended Section 123 of the AEA to require the President to keep the Senate Foreign Relations Committee and the House Foreign Affairs Committee "fully and currently informed of any initiative or negotiations relating to a new or amended agreement for peaceful nuclear cooperation."

Export Licensing

The AEA sets out procedures for licensing exports to states with which the United States has nuclear cooperation agreements. (Sections 126, 127, and 128 codified as amended at 42 U.S.C. 2155, 2156, 2157.) Each export of nuclear material, equipment, or technology requires a specific export license or other authorization. The Nuclear Regulatory Commission (NRC) is required to meet criteria in Sections 127 and 128 in authorizing export licenses. These criteria are as follows:

- Application of IAEA safeguards to any material or facilities proposed to be exported, material or facilities previously exported, and to any special nuclear material used in or produced through the use thereof (these are not full-scope safeguards, but safeguards required under Article III.2 of the nuclear Nonproliferation Treaty [NPT]).

- Nothing exported can be used for any nuclear explosive device or for research on or development of any nuclear explosive device.
- Recipient states must have adequate physical security on "such material or facilities proposed to be exported and to any special nuclear material used in or produced through the use thereof."
- Recipient states are not to retransfer exported nuclear materials, facilities, sensitive nuclear technology, or "special nuclear material produced through the use of such material" without prior U.S. approval.
- Recipient states may not reprocess or alter in form or content exported nuclear material or special nuclear material produced though the use of exported nuclear material without prior U.S. approval.
- The foregoing conditions must be applied to any nuclear material or equipment that is produced or constructed under the jurisdiction of the recipient by or through the use of any exported sensitive nuclear technology.
- Section 128 requires that recipient non-nuclear weapon states must have full-scope IAEA safeguards.

The President must judge that the proposed export or exemption will "not be inimical to the common defense and security" or that any export of that type "would not be inimical to the common defense and security because it lacks significance for nuclear explosive purposes." The executive branch may also consider other factors, such as "whether the license or exemption will materially advance the nonproliferation policy of the United States by encouraging the recipient nation to adhere" to the NPT; whether "failure to issue the license or grant the exemption would otherwise be seriously prejudicial" to U.S. nonproliferation objectives; and whether the recipient nation has agreed to conditions identical to those laid out in Section 127.

Section 126b.(2) contains a provision for the President to authorize an export in the event that the NRC deems that the export would not meet Section 127 and 128 criteria. The President must determine "that failure to approve an export would be seriously prejudicial to the achievement of U.S. nonproliferation objectives or otherwise jeopardize the common defense and security." In that case, the President would submit his executive order, along with a detailed assessment and other documentation, to Congress for 60 days of continuous session. After 60 days of continuous session, the export would go through unless Congress were to adopt a concurrent resolution of disapproval.[9]

Section 128b.(2) contains a provision for the President to waive termination of exports by notifying Congress that the state has adopted full-scope safeguards or that the state has made significant progress toward adopting such safeguards, or that U.S. foreign policy interests dictate reconsideration. Such a determination would become effective unless Congress were to adopt a concurrent resolution of disapproval within 60 days of continuous session.

Additionally, Section 129b.(1) forbids the export of "nuclear materials and equipment or sensitive nuclear technology" to any country designated as a state sponsor of terrorism.[10] Section 129b.(3) allows the President to waive this provision.

Iran-Related Restrictions

The Comprehensive Iran Sanctions, Accountability, and Divestment Act (CISADA) of 2010 (P.L. 111-195), which became law July 1, 2010,

---

[9] In light of the Supreme Court's 1983 decision in *INS v. Chadha*, passing a concurrent resolution could invite a legal challenge because it is arguably unconstitutional. Although not provided for in the AEA, Congress could choose to pass a joint resolution of disapproval or a bill stating in substance it did not approve.

[10] Section 129b. (2) states that the prohibitions described in the previous section "shall not apply to exports, reexports, transfers, or retransfers of radiation monitoring technologies, surveillance equipment, seals, cameras, tamper-indication devices, nuclear detectors, monitoring systems, or equipment necessary to safely store, transport, or remove hazardous materials ... except to the extent that such technologies, equipment, seals, cameras, devices, detectors, or systems are available for use in the design or construction of nuclear reactors or nuclear weapons."

contains additional restrictions on licensing nuclear exports to countries with entities that have been sanctioned for conducting certain types of nuclear weapons-related transactions with Iran. Section 102a.(2)(A) of the law states that "no license may be issued for the export, and no approval may be given for the transfer or retransfer" of "any nuclear material, facilities, components, or other goods, services, or technology that are or would be subject to an agreement for cooperation between the United States" and such countries. Section 102 a.(2)(B), however, allows the President to waive these restrictions. Section 102a.(2)(C) allows the President to authorize licenses for nuclear exports "on a case-by-case basis" to entities (which have not been sanctioned) in countries subject to the restrictions described above.[11]

SUBSEQUENT ARRANGEMENTS

Section 131 of the AEA details procedures for subsequent arrangements to nuclear cooperation agreements concluded pursuant to Section 123. Such arrangements are required for forms of nuclear cooperation requiring additional congressional approval, such as transfers of nuclear material or technology and recipient states' enrichment or reprocessing of nuclear materials transferred pursuant to the agreement. Subsequent arrangements may also include arrangements for physical security, storage, or disposition of spent nuclear fuel; the application of safeguards on nuclear materials or equipment; or "any other arrangement which the President finds to be important from the standpoint of preventing proliferation."

Before entering into a subsequent arrangement, the Secretary of Energy must publish in the *Federal Register* a determination that the arrangement "will not be inimical to the common defense and security." A proposed subsequent arrangement shall not take effect before 15 days after publication of both this determination and notice of the proposed

[11] For details on these sanctions, see CRS Report RS20871, *Iran Sanctions*, by Kenneth Katzman.

arrangement. The Secretary of State is required to prepare an unclassified Nuclear Proliferation Assessment Statement (NPAS) if, "in the view of" the Secretary of State, Secretary of Energy, Secretary of Defense, or the Nuclear Regulatory Commission, a proposed subsequent arrangement "might significantly contribute to proliferation." The Secretary of State is to submit the NPAS to the Secretary of Energy within 60 days of receiving a copy of the proposed subsequent arrangement. The President may waive the 60-day requirement if the Secretary of State so requests, but must notify both the House Foreign Affairs Committee and Senate Foreign Relations Committee of any such waiver and the justification for it. The Secretary of Energy may not enter into the subsequent arrangement before receiving the NPAS.

Section 131 specifies requirements for certain types of subsequent arrangements. Section 131b. describes procedures for the executive branch to follow before entering into a subsequent arrangement involving the reprocessing of U.S.-origin nuclear material or nuclear material produced with U.S.-supplied nuclear technology. These procedures also cover subsequent arrangements allowing the retransfer of such material to a "third country for reprocessing" or "the subsequent retransfer" of more than 500 grams of any plutonium produced by reprocessing such material. The Secretary of Energy must provide both the House Foreign Affairs Committee and Senate Foreign Relations Committee with a report describing the reasons for entering into the arrangement. Additionally, 15 days of continuous session must elapse before the Secretary may enter into the arrangement, unless the President judges that "an emergency exists due to unforeseen circumstances requiring immediate entry" into the arrangement. In such a case, the waiting period would be 15 calendar days.

If a subsequent arrangement described in the above paragraph involves a facility that has not processed spent nuclear reactor fuel prior to March 10, 1978 (when the Nuclear Nonproliferation Act of 1978 was enacted), the Secretaries of State and Energy must judge that the arrangement "will not result in a significant increase of the risk of proliferation." In making this judgment, the Secretaries are to give "foremost consideration ... to whether or not the reprocessing or retransfer will take place under conditions that will ensure timely warning to the United States of any diversion well in advance of the time at which the non-nuclear weapon state could transform the diverted material into a nuclear explosive device."[12] For a subsequent arrangement involving reprocessing in a facility that has processed spent nuclear reactor fuel prior to March 10, 1978, the Secretary of Energy will "attempt to ensure" that reprocessing "shall take place under conditions" that would satisfy the timely-warning conditions described above. Section 131f. specifies procedures for congressional approval of subsequent arrangements involving the storage or disposition of foreign spent nuclear fuel in the United States.

Section 133 states that, before approving a subsequent arrangement involving certain transfers of special nuclear material, the Secretary of Energy must consult with the Secretary of Defense "on whether the physical protection of that material during the export or transfer will be adequate to deter theft, sabotage, and other acts of international terrorism which would result in the diversion of that material."[13] If the Secretary of Defense determines that "the export or transfer might be subject to a genuine terrorist threat," that Secretary is required to provide a written risk assessment of the risk and a "description of the actions" that he or she "considers necessary to upgrade physical protection measures."

Examples of Subsequent Arrangements

U.S.-Japan Agreement

The first test of the subsequent arrangement provisions came in August 1978, when the Department of Energy informed the House and Senate foreign relations committees of a Japanese request for approval of the transfer of spent fuel assemblies from Japan to the United Kingdom for reprocessing. This was the first "subsequent arrangement" approved. The

---

[12] These provisions also apply to facilities that, prior to March 18, 1978, did not have a subsequent arrangement for reprocessing.

[13] This section applies to "the export or transfer of more than 2 kilograms of plutonium or more than 5 kilograms of uranium enriched to more than 20 percent in the isotope 233 or the isotope 235."

United States and Japan entered into similar arrangements until 1988, when the two governments revised their nuclear cooperation agreement. That agreement included an "implementing agreement," which provided 30-year advance consent for the transfer of spent fuel from Japan to Europe for reprocessing. While controversial, Congress did not block the nuclear cooperation agreement.

A subsequent arrangement was also necessary for the sea transport from Europe to Japan of plutonium that had been separated from the Japanese spent fuel. The Department of Energy approved a Japanese request for 30-year advance consent for the sea transport of plutonium. It was submitted to Congress as a subsequent arrangement, and took effect in October 1988.

U.S.-India Agreement

The U.S. nuclear cooperation agreement with India grants New Delhi consent to reprocess nuclear material transferred pursuant to the agreement, as well as "nuclear material and byproduct material used in or produced through the use of nuclear material, non-nuclear material, or equipment so transferred." However, the agreement also includes a requirement that India first build a new national reprocessing facility to be operated under IAEA safeguards. The two countries signed a subsequent arrangement July 30, 2010, which governs the procedures for operating two new reprocessing facilities in India. The agreement also describes procedures for U.S. officials to inspect and receive information about physical protection measures at the new facilities. The arrangement would not have taken effect if Congress had adopted a joint resolution of disapproval within 30 days of continuous session; Congress did not adopt such a resolution.[14]

---

[14] Section 201 of the United States-India Nuclear Cooperation Approval and Nonproliferation Enhancement Act (P.L. 110-369), which approved the U.S.-India cooperation agreement, specifies procedures (different from those described in the Atomic Energy Act) for Congress to consider subsequent arrangements to that agreement.

If India were to construct any additional facilities to reprocess fuel from U.S.-supplied reactors, a new subsequent arrangement would need to be submitted to Congress.

TERMINATION OF COOPERATION

Section 129a. of the AEA requires that the United States end exports of nuclear materials and equipment or sensitive nuclear technology to any non-nuclear weapon state that, after March 10, 1978, the President determines to have detonated a nuclear explosive device; terminated or abrogated IAEA safeguards; materially violated an IAEA safeguards agreement; or engaged in activities involving source or special nuclear material and having "direct significance" for the manufacture or acquisition of nuclear explosive devices, and "has failed to take steps which, in the President's judgment, represent sufficient progress toward terminating such activities."

Section 129a. also requires that the United States halt exports to any nation the President determines to have materially violated the terms of an agreement for cooperation with the United States; assisted, encouraged, or induced any non-nuclear weapon state to obtain nuclear explosives or the materials and technologies needed to manufacture them; or re-transferred or entered into an agreement for exporting reprocessing equipment, materials, or technology to a non-nuclear weapon state, unless in connection with an international agreement to which the United States subscribes.

The President can waive termination of exports if he determines that "cessation of such exports would be seriously prejudicial to the achievement of United States nonproliferation objectives or otherwise jeopardize the common defense and security." The President must submit his determination to Congress, which is then referred to the House Committee on Foreign Affairs and the Senate Foreign Relations Committee for 60 days of continuous session. The determination becomes effective unless Congress adopts a joint resolution opposing the determination.

PART 810 AGREEMENTS

Section 57.b. (2) of the Atomic Energy Act allows for limited forms of nuclear cooperation related to the "development or production of any special nuclear material outside of the United States" if that activity has been authorized by the Secretary of Energy following a determination that it "will not be inimical to the interest of the United States." The Secretary may only make such a finding with "the concurrence of the Department of State, and after consultation with the Nuclear Regulatory Commission [NRC], the Department of Commerce, and the Department of Defense."[15] Authorizations of such activities are also known as "Part 810 authorizations," after 10 Code of Federal Regulations (C.F.R.) Part 810. Part 810 regulations describe activities that are "generally authorized" by the Secretary of Energy and activities that require "specific authorization" by the Secretary. Some "generally authorized activities" are limited to a list of "generally authorized destinations."[16] These regulations also detail "reporting requirements for authorized activities."

Part 810 authorizations mostly involve unclassified nuclear technology transfer and services, such as nuclear reactor designs, nuclear facility operational information and training, and nuclear fuel fabrication. Such an authorization is not required for exports of components and materials licensed by NRC governed by 10 C.F.R. Part 110. Civilian nuclear cooperation agreements under Section 123 of the Atomic Energy Act of 1954, as amended (hereinafter Atomic Energy Act or AEA), are not required for an 810 authorization or for transmission of nuclear-related information, except for restricted data. Such agreements are, however, required for such forms of nuclear cooperation as the transfer of U.S.-origin special nuclear material subject to licensing for commercial, medical, and industrial purposes; the export of reactors and critical parts of reactors; and other commodities under NRC export licensing authority (10

---

[15] The Secretary must also consult the Office of the Director of National Intelligence in the case of transfers to China or Russia.

[16] A list of such destinations is available at https://www.law.cornell.edu/cfr/text/10/appendix-A_to_part_810.

C.F.R. 110).[17] The NRC may also authorize activities governed by Part 810 authorizations under a 123 agreement or under a subsequent arrangement to such an agreement.

It is worth noting that Part 810.9 includes "[w]hether the United States has an agreement for cooperation in force covering exports to the country or entity involved" as a factor for the Secretary of Energy to use in determining that an activity "will not be inimical to the interest [sic] of the United States." Moreover, the list of "generally authorized destinations" is "based principally on the United States agreements for civil nuclear cooperation," according to guidance from the National Nuclear Security Administration.[18]

Recent Legislative Activity

In recent years, Congress has intensely debated the requirements of the Atomic Energy Act for nuclear cooperation agreements, most notably following the approval of the 2008 U.S.-India nuclear cooperation agreement. Congress has held several hearings to discuss whether stronger nonproliferation standards and congressional oversight for nuclear cooperation agreements are desirable.

On March 21, 2018, Representative Ileana Ros-Lehtinen introduced H.R. 5357, the Nuclear Cooperation Reform Act of 2018, which would amend the Atomic Energy Act to require nonexempt nuclear cooperation agreements to include several additional provisions. These provisions include a legally binding "commitment" from the cooperating government stipulating that "no enrichment or reprocessing activities, or acquisition or construction of such facilities, [would] occur within the territory over which the cooperating party exercises sovereignty"; "a guaranty by the cooperating party that no nationals of a third country" would be "permitted

[17] Lists of exports governed by this authority are available at https://www.nrc.gov/about-nrc/ip/export-import.html.

[18] https://nnsa.energy.gov/sites/default/files/nnsa/inlinefiles/nei_faqs_final_9-12-16_final_gc-53_adh.pdf.

access to any reactor, related equipment, or sensitive materials transferred under" the agreement without prior U.S. consent; and a "commitment to maintain" or enact "a legal regime providing for adequate protection from civil liability that will allow for the participation of United States suppliers in any effort by the country to develop civilian nuclear power."

The bill would also require a cooperating party to sign, ratify, and implement an additional protocol to its IAEA safeguards agreement; implement a number of export control-related measures specified by the bill; comply with "all United Nations conventions to which the United States is a party and all [UN] Security Council resolutions regarding the prevention of the proliferation of weapons of mass destruction"; and be party to, as well as fully implement, "the provisions and guidelines" of the Biological Weapons Convention and the Chemical Weapons Convention, as well as "all other international agreements to which the United States is a party regarding the export of nuclear, chemical, biological, and advanced conventional weapons, including missiles and other delivery systems."

H.R. 5357 contains provisions concerning U.S. foreign assistance. For example, the bill would prohibit "assistance (other than humanitarian assistance) under any provision of law ... to a country that has withdrawn" from the NPT. The United States would be required to "seek the return of any material, equipment, or components transferred under" a nuclear cooperation agreement. In addition, the bill would prohibit any assistance

> under the Foreign Assistance Act of 1961, the Arms Export Control Act, the Foreign Military Sales Act, the Food for Peace Act, the Peace Corps Act, or the Export-Import Bank Act of 1945 to any country if the Secretary of State determines that the government of the country has repeatedly provided support for acts of proliferation of equipment, technology, or materials to support the design, acquisition, manufacture, or use of weapons of mass destruction or the acquisition or development of missiles to carry such weapons.

This section of the bill includes a presidential waiver provision. H.R. 5357 would also permit joint resolutions approving nuclear cooperation agreements to "include any other provisions to accompany such proposed agreement for cooperation."

Moreover, H.R. 5357 amends the AEA's congressional notification provisions concerning ongoing nuclear cooperation agreement negotiations. Specifically, the bill requires the President to "consult" with the Senate Foreign Relations Committee and the House Foreign Affairs Committee "concerning such initiative or negotiations beginning not later than 15 calendar days after the initiation of any such negotiations, or the receipt or transmission of a draft agreement, whichever occurs first, and monthly thereafter until such time as the negotiations are concluded." These consultations are to include the provision of "current working drafts and proposed text put forward for negotiation by the parties for inclusion in such agreement."

Appendix A. Key Dates for Bilateral Civilian Nuclear Cooperation ("Section 123") Agreements

Table A-1. Key Dates

Country	Most Recent Agreement Signed	Entered into Force	Duration	Expiration	Renewal Terms[a]	Original Agreement[b]
Argentina	February 29, 1996	October 16, 1997	30 years	October 16, 2027	Extension by agreement of the parties	
Australia	May 4, 2010	December 22, 2010	30 years	December 22, 2040	Automatic 5-yr renewals after 30 years	1956
Brazil	October 14, 1997	September 15, 1999	30 years	September 15, 2029	Extension by agreement of the parties	1972
Canada	June 23, 1999	December 13, 1999	30 years	January 1, 2030	Automatic 5-yr renewals after 30 years	1955
China[c]	April 13, 2015	November 10, 2015	30 years	2045	None specified	1985
Egypt	June 29, 1981	December 29, 1981	40 years	December 29, 2021	None specified	

Country	Most Recent Agreement Signed	Entered into Force	Duration	Expiration	Renewal Terms[a]	Original Agreement[b]
European Atomic Energy Community (Euratom)[d]	November 7, 1995	March 29, 1996	30 years	March 29, 2026	Automatic 5-yr renewals after 30 years	1958
India[e]	October 10, 2008	December 6, 2008	40 years	December 6, 2048	Automatic 10-yr renewals after 40 years	
Indonesia	June 30, 1980	December 30, 1981	50 years	December 30, 2031	None specified	1960
International Atomic Energy Agency (IAEA)	May 11, 1959	August 7, 1959	95 years (Amended in 1974, 1980, Renewed in 2014)	August 7, 2054	None specified	1959
Japan	November 4, 1987	July 17, 1988	30 years	July 17, 2018	Remains in force until terminated by a party	1968
Kazakhstan	November 18, 1997	November 5, 1999	30 years	November 5, 2029	None specified	
Republic of Korea	November 24, 1972 [proposed renewal agreement signed June 15, 2015]	November 25, 2015	20 years	November 24, 2035	Automatic 5-yr renewals after 20 years	1956
Morocco	May 30, 1980	May 16, 1981	30 years	May 16, 2021	Automatic 5-yr renewals after 30 years	
Norway[f]	June 11, 2016		30 years	Thirty years after entry into force	None specified	1984
Russian Federation	May 6, 2008	January 11, 2011	30 years	January 11, 2041	None specified	
South Africa	August 25, 1995	December 4, 1997	25 years	December 4, 2022	None specified	1957
Switzerland	October 31, 1997	June 23, 1998	30 years	June 23, 2028	Automatic 5-yr renewals after 30 years	1965

Table A-1. (Continued)

Country	Most Recent Agreement Signed	Entered into Force	Duration	Expiration	Renewal Terms[a]	Original Agreement[b]
Taiwan (TECRO)	December 20, 2013	June 22, 2014	amended in 1974, to 30 years; renewed in 2014 to indefinite duration	none	n/a	1955
Turkey	July 26, 2000	June 2, 2008	15 years	June 2, 2023	Automatic 5-yr renewals after 15 years	
Ukraine	May 6, 1998	May 28, 1999	30 years	May 28, 2029	None specified	
United Arab Emirates	May 21, 2009	December 17, 2009	30 years	December 17, 2039	None specified	
Vietnam	May 6, 2014	October 3, 2014	30 years	October 3, 2044	Automatic 5-yr renewals after 30 years	

Sources: CRS; Text of Agreements; U.S. Department of State Fact Sheet, "U.S. Bilateral Agreements for Peaceful Nuclear Cooperation Pursuant to Section 123 of the U.S. Atomic Energy Act of 1954, as amended," December 5, 2013.

[a] If renewal terms are not specified, then a new Section 123 agreement would need to be negotiated and submitted to Congress for the required review.

[b] The "Original Agreement" field refers to the year that the first civilian nuclear cooperation agreement was concluded with that country. If it is blank, the current agreement is the first such agreement.

[c] P.L. 99-183 approved the original 1985 agreement but prohibited licenses from being issued until the President certified that transferred items would be used for solely peaceful purposes and reported to Congress on China's nonproliferation policies. Following the Tiananmen Square crackdown, P.L. 101-246 (FY1990 Foreign Relations Authorization Act) also suspended nuclear cooperation with China. President Clinton issued the required waiver, report, and certification in January 1998. The required congressional review period ended on March 18, 1998, and implementation of the agreement was then allowed. See also CRS Report RL33192, U.S.-China Nuclear Cooperation Agreement, by Mark Holt, Mary Beth D. Nikitin, and Paul K. Kerr.

[d] Euratom member states include Austria, Belgium, Bulgaria, Croatia, Cyprus, Czech Republic, Denmark, Estonia, Finland, France, Germany, Greece, Hungary, Ireland, Italy, Latvia, Lithuania, Luxembourg, Malta, the Netherlands, Poland, Portugal, Romania, Slovakia, Slovenia, Spain, Sweden, and the United Kingdom.

[e] P.L. 109-401 and P.L. 110-369 approved the agreement with conditions. See CRS Report RL33016, *U.S. Nuclear Cooperation with India: Issues for Congress*, by Paul K. Kerr.

[f] P.L. 114-320 approved the 2016 renewal agreement.

APPENDIX B. ENRICHMENT AND REPROCESSING RESTRICTIONS

Although some experts have advocated requiring governments to forgo enrichment and reprocessing (a nonproliferation commitment sometimes referred to as the "Gold Standard") as a condition for concluding a nuclear cooperation agreement, the Atomic Energy Act (AEA) does not include such a requirement. In recent years, the United States has attempted to persuade certain countries with which it is negotiating nuclear cooperation agreements to forgo enrichment and reprocessing and conclude additional protocols to their International Atomic Energy Agency (IAEA) safeguards agreements; past U.S. nuclear cooperation agreements have not included these additional components. The AEA does mandate that U.S. nuclear cooperation agreements require U.S. consent for any "alteration in form or content" (to include enrichment or reprocessing) of U.S.-origin material or any material processed in a plant containing transferred U.S. nuclear technology. Such agreements also require U.S. consent for any retransfer of material or technology.

The United States has argued that its December 2009 nuclear cooperation agreement with the United Arab Emirates (UAE) could set a useful precedent for mitigating the dangers of nuclear proliferation.[19] For example, President Barack Obama's May 21, 2009, letter transmitting the agreement to Congress argued that the agreement had "the potential to serve as a model for other countries in the region that wish to pursue responsible nuclear energy development." Similarly, then-State Department spokesperson P.J. Crowley described the agreement as "the gold standard" during an August 5, 2010, press briefing, although the Obama Administration generally did not use this term when describing its nuclear cooperation policies.

---

[19] For more information, see CRS Report R40344, *The United Arab Emirates Nuclear Program and Proposed U.S. Nuclear Cooperation*, by Christopher M. Blanchard and Paul K. Kerr.

The U.S.-UAE agreement's status as a potential model is grounded in two nonproliferation provisions not found in other U.S. nuclear cooperation agreements. First, the agreement requires the country to bring into force the additional protocol to its safeguards agreement before the United States licenses "exports of nuclear material, equipment, components, or technology" pursuant to the agreement.[20] Second, the agreement states that the UAE

> shall not possess sensitive nuclear facilities within its territory or otherwise engage in activities within its territory for, or relating to, the enrichment or reprocessing of material, or for the alteration in form or content (except by irradiation or further irradiation or, if agreed by the Parties, post-irradiation examination) of plutonium, uranium 233, high enriched uranium, or irradiated source or special fissionable material.

The U.S.-UAE agreement also provides the United States with the right to terminate nuclear cooperation and to require the return of any nuclear "material, equipment or components ... and any special fissionable material produced through their use" if, after the agreement's entry into force, the UAE "possesses sensitive nuclear facilities within its territory or otherwise engages in activities within its territory relating to enrichment of uranium or reprocessing of nuclear fuel."[21]

Notwithstanding its characterization of the U.S.-UAE agreement, the Obama Administration announced in December 2013 after an interagency review that renouncing domestic enrichment and reprocessing would not be a prerequisite to concluding a nuclear cooperation agreement for all countries, and each partner country would be considered individually.[22] The U.S. nuclear cooperation agreement with Vietnam, which the two governments concluded in 2014, did not include a provision requiring the

---

[20] The IAEA Board of Governors approved the Protocol March 3, 2009. The UAE signed it the next month, and brought it into force December 20, 2010.

[21] The AEA requires that there is no enrichment or reprocessing by the recipient state of transferred nuclear material or nuclear material produced with materials or facilities transferred pursuant to the agreement without prior approval.

[22] Daniel Horner, "U.S. Policy of Nuclear Pacts Detailed," *Arms Control Today*, January/ February 2014.

country to forgo enrichment and reprocessing, although the agreement's preamble includes a political commitment stating that Vietnam intends to rely on international markets for its nuclear fuel supply, rather than acquiring sensitive nuclear technologies.[23]

APPENDIX C. NUCLEAR COOPERATION AGREEMENTS APPROVED OUTSIDE ATOMIC ENERGY ACT PROCESS

Congress has used legislation to approve nuclear cooperation agreements that did not use the legislative process mandated by the Atomic Energy Act (AEA) of 1954, as amended.

Australia[24]

On May 5, 2010, President Barack Obama submitted a renewed U.S.-Australia nuclear cooperation agreement to Congress for approval. H.R. 6411, which the House adopted on November 30, 2010, would have approved the agreement even if there had not been sufficient legislative days remaining in the 111th Congress; the Senate did not adopt its version of the bill (S. 3844). These bills were not needed because the 111th Congress contained a sufficient number of days for the agreement to enter into force.

China[25]

In 1985, President Ronald Reagan submitted the first U.S.-China nuclear cooperation agreement to Congress, which adopted a joint

[23] For more information about the Vietnam agreement, see CRS Report R43433, *U.S.-Vietnam Nuclear Cooperation Agreement: Issues for Congress*, by Mary Beth D. Nikitin, Mark Holt, and Mark E. Manyin.

[24] For more information, see CRS Report R41312, *U.S.-Australia Civilian Nuclear Cooperation: Issues for Congress*, by Mary Beth D. Nikitin and Bruce Vaughn.

[25] For more information, see CRS Report RL33192, *U.S.-China Nuclear Cooperation Agreement*, by Mark Holt, Mary Beth D. Nikitin, and Paul K. Kerr.

resolution, P.L. 99-183, requiring that the President make certain nonproliferation-related certifications in order for the agreement to be implemented. P.L. 99-183 required a presidential certification and a report followed by a period of 30 days of continuous session of Congress. P.L. 101-246, the Foreign Relations Authorization Act for Fiscal Years 1990 and 1991, imposed sanctions on China, including suspending nuclear cooperation and requiring an additional presidential certification on Beijing's nuclear nonproliferation assurances. Before a summit with China, President William Clinton on January 12, 1998, signed the required certifications regarding China's nuclear nonproliferation policy and practices. Clinton also issued a certification and waived a sanction imposed under P.L. 101-246. Congressional review ended on March 18, 1998, allowing the agreement to be implemented.

India[26]

P.L. 109-401, which became law on December 18, 2006, permitted the President to waive several provisions of the AEA with respect to a nuclear cooperation agreement with India. On September 10, 2008, President George W. Bush submitted to Congress a determination that P.L. 109-401's requirements for such an agreement to proceed had been met. President Bush signed P.L. 110- 369, which approved the agreement, into law on October 8, 2008.

Norway

The President submitted an extension of the U.S.-Norway nuclear cooperation agreement to Congress on June 14, 2016. P.L. 114-320, which became law on December 16, 2016, approved the agreement

[26] For more information, see CRS Report RL33016, *U.S. Nuclear Cooperation with India: Issues for Congress*, by Paul K. Kerr.

"[n]otwithstanding the provisions for congressional consideration" in the AEA, thereby addressing concerns that that there was an insufficient number of legislative days remaining in the 114th Congress for congressional consideration.

In: Energy Policies: Issues and Legislation ISBN: 978-1-53614-485-7
Editor: Doug Higgins

Chapter 4

NERC Standards for Bulk Power Physical Security: Is the Grid More Secure?*

Paul W. Parfomak

Summary

A 2013 rifle attack on a critical electric power substation in Metcalf, CA, marked a turning point for the U.S. electric power sector. The attack prompted utilities across the country to reevaluate and restructure their physical security programs. It also set in motion proceedings in Congress and at the Federal Energy Regulatory Commission (FERC) which resulted in a new mandatory Physical Security Reliability Standard (CIP-014) for bulk power asset owners promulgated by the North American Electric Reliability Corporation (NERC) in 2015. In the three years since FERC approved this new standard, security risks to the power grid have become an even greater concern in the electric utility industry. Reflecting these ongoing security concerns, legislative proposals in the 115th Congress include provisions directed at power grid physical security.

* This is an edited, reformatted and augmented version of a Congressional Research Service report, R45135, dated March 19, 2018.

Congress also continues its oversight of grid security and implementation of NERC's security standards.

Three entities play key roles in standards oversight and support of implementation for bulk power physical security. NERC and FERC oversee implementation of the CIP-014 standards, while the Department of Energy plays a supporting role in helping bulk power asset owners to protect their critical infrastructure. The detailed findings of NERC's compliance activities are not publicly disclosed due to their confidential nature. However, NERC has stated that the utility industry is making progress towards effective implementation of the CIP-014 standard and NERC has been "encouraged" by grid security measures put in place so far. NERC compliance audits as of February 2018 have uncovered no major failures to date.

In addition to compliance with NERC's standards, there have been other observable changes within the electricity sector reflecting greater emphasis on bulk power physical security. These changes include realignment in corporate structure to support physical security, incorporating physical security in transmission planning, new security products and services, utility capital investment in physical security, and utility participation in voluntary security programs. While public information about such changes is limited, it suggests they may be significant and widespread.

Although the electric power sector seems to be moving in the overall direction of greater physical security for critical assets, many measures have yet to be implemented and the process of corporate realignment around physical security is still underway. NERC's CIP-014 standards have been promulgated recently, and bulk power asset owners have largely begun enhancing physical security under the standard over the last two years. Therefore, although it is probably accurate to conclude that, based on the objectives of the CIP-014 standards, the U.S. electric grid is more physically secure than it was five years ago, it has not necessarily reached the level of physical security needed based on the sector's own assessments of risk. Bulk power security remains a work in progress.

Congress continues to be concerned about the current state of electric grid physical security. Among many specific issues of potential interest, Congress may focus on several with policy significance: security implementation oversight, cost recovery, hardening vs. resilience, and the quality of threat information. As CIP-014 implementation and other physical security initiatives proceed, Congress also may seek to maintain its focus on the power sector's overall progress, not only on short term compliance with NERC's security standards, but also on structural changes supporting physical security as a priority far into the future.

INTRODUCTION

Securing the electric power grid is among the highest priorities for critical infrastructure protection in the United States. In the past, power grid facilities have had varying degrees of access control and surveillance depending upon the facility type and location. These measures were largely focused on public safety (reflecting liability concerns) and preventing vandalism and theft. More recently, federal agencies, Congress, and the utility industry have focused greater attention on the vulnerability of the power grid, especially the high voltage transmission (bulk power) system, to terrorist attacks which could cause widespread, extended blackouts.

Until 2013, the emphasis of analysts and policymakers was on power grid cybersecurity— protecting the computer controls and communication systems used to operate the grid. However, a 2013 rifle attack on an electric transmission substation in Metcalf, CA, shifted more attention to the physical security of power grid critical assets. In response to the Metcalf attack, as well as other grid incidents and findings from utility security exercises, Congress passed new legislation to strengthen power grid physical security and to facilitate recovery in the event of a successful attack.[1] Congress also sought stronger physical security standards from the Federal Energy Regulatory Commission (FERC) under the commission's existing statutory authority to regulate the reliability of the bulk power system. FERC, in turn, ordered the North American Electric Reliability Corporation (NERC)—the not-for-profit organization responsible for ensuring grid reliability—to promulgate new requirements for the physical security of bulk power critical infrastructure.[2] After consultation within the utility industry, NERC proposed new physical security standards in May

[1] The Fixing America's Surface Transportation (FAST) Act (P.L. 114-94), which became law on December 4, 2015, contains provisions to protect or restore the reliability of critical electric infrastructure or defense of critical electric infrastructure during a grid security emergency (§1104).

[2] Among other functions, NERC develops and enforces reliability standards, monitors the grid, and trains industry personnel. In the United States, NERC is subject to Federal Energy Regulatory Commission oversight.

2014. FERC approved them, with minor changes, the following November.[3]

Since 2014, security risks to the power grid have become an even greater concern in the electric utility industry. Addressing them has remained a concern of Congress.[4] An emphasis on physical risk to the power grid was underscored in September 2016 by another successful rifle attack on a transformer substation—in Utah. Reflecting ongoing security concerns, legislative proposals in the 115th Congress include provisions directed at power grid physical security. Congress also continues its oversight of FERC's grid security activities and the implementation of NERC's physical security standards.

This chapter examines changes to the physical security of the electric power grid since the promulgation of NERC's physical security standards. The chapter discusses the current risk environment for the bulk power system. It summarizes the key requirements of NERC's security standards, including its applicability to specific assets, implementation deadlines, and oversight. The chapter reviews observable changes in the utility sector related to physical security. It concludes with an overview of proposed legislation and a discussion of policy issues for Congress.

This chapter focuses primarily on physical security efforts to prevent successful physical attacks on the bulk power system. This chapter also does not address issues related to security incident recovery or restoration, except in the context of preventive physical security.

POWER GRID THREAT ENVIRONMENT

Grid security analysts and policymakers have long been aware of physical risks to bulk power critical infrastructure, especially to high

---

[3] For more historical background and details regarding the development of NERC's standards, see CRS Report R43604, *Physical Security of the U.S. Power Grid: High-Voltage Transformer Substations*, by Paul W. Parfomak.

[4] See for example: Senator Ron Johnson, Chairman, Opening statement before the Senate Committee on Homeland Security and Governmental Affairs hearing on "Threats to the Homeland," September 27, 2017.

voltage (HV) transformer stations and substations, which serve as key nodes within the electric transmission system.[5] The 2013 Metcalf attack, in which an unknown perpetrator firing a .30 caliber rifle disabled a critical 500 kilovolt (kV) transformer substation, demonstrated that such facilities face real and potentially sophisticated threats.[6] The September 2016 rifle attack on a 69 kV transformer substation in Utah—which reportedly left 13,000 rural customers without power for up to eight hours—showed that similar incidents could occur almost anywhere on the grid.[7] A successful cyberattack on Ukraine's power grid in 2015, which was reportedly attributed to Russian hackers, showed that foreign entities could view power grids as attractive targets.[8] A 2017 report from the National Academy of Sciences concludes: "While to date there have been only minor attacks on the power system in the United States, large-scale physical destruction of key parts of the power system by terrorists is a real danger. Some physical attacks could cause disruption in system operations that last for weeks or months."[9]

The persistent threat environment has been changing the perception of physical threats among power grid owners and operators. For example, surveys of electric utility employees show that their physical (and cyber) security concerns are growing.[10] Exelon Corporation, one of the nation's largest utility holding companies, stated in its 2016 annual report:

> Threat sources continue to seek to exploit potential vulnerabilities in the electric...utility industry associated with protection of sensitive and

---

[5] See, for example: National Research Council, Terrorism and the Electric Power Delivery System, 2012 and Office of Technology Assessment, Physical Vulnerability of Electric Systems to Natural Disasters and Sabotage, OTA-E-453, June 1990.

[6] RTO Insider, "Substation Saboteurs 'No Amateurs,'" April 2, 2014, https://www.rtoinsider.com/pjm-grid2020-1113-03/

[7] Pat Reavy, "Power Company Offers Rare $50K Reward for Information on Vandalism," Deseret News, September 29, 2016. A substation rated at 69 kilovolts is not considered a "high voltage" transmission asset, although it may still serve large numbers of customers.

[8] Jim Finkle, "U.S. Firm Blames Russian 'Sandworm' Hackers for Ukraine Outage," Reuters, January 7, 2016. The attack reportedly cut power to 80,000 customers for about six hours.

[9] National Academy of Sciences, Engineering, and Medicine, Enhancing the Resilience of the Nation's Electricity System, 2017, p. 65, https://doi.org/10.17226/24836.

[10] Utility DIVE, *2017 State of the Electric Utility Survey*, April 10, 2017, https://s3.amazonaws.com/dive_assets/rlpsys/SEU_2017.pdf

> confidential information, grid infrastructure and other energy infrastructures, and such attacks and disruptions, both physical and cyber, are becoming increasingly sophisticated and dynamic....The risk of these system-related events and security breaches occurring continues to intensify....[11]

Xcel Energy, another major utility owner, likewise states in its 2016 annual report:

> Our generation plants, fuel storage facilities, transmission and distribution facilities and information systems may be targets of terrorist activities... The potential for terrorism has subjected our operations to increased risks and could have a material effect on our business.[12]

Accordingly, electricity sector-wide security exercises conducted by NERC have simulated attacks on power grid critical assets combining both cyber and physical dimensions.[13] These exercises are further discussed later in this chapter.

NERC's Physical Security Standards

On March 7, 2014, FERC ordered NERC to submit proposed reliability standards requiring transmission owners meeting certain criteria "to take steps or demonstrate that they have taken steps to address physical security risks and vulnerabilities related to the reliable operation" of the power grid.[14] In its order FERC stated that physical security standards were

---

[11] Exelon Corporation, *Annual Report Pursuant to Section 13 or 15(d) of the Securities and Exchange Act of 1934 for the Fiscal Year Ended December 31, 2016*, Form 10-K, February 13, 2017, p. 63.

[12] Excel Energy, Inc. *Annual Report Pursuant to Section 13 or 15(d) of the Securities Exchange Act of 1934 for the Fiscal Year Ended December 31, 2016*, Form 10-K, p. 44.

[13] North American Electric Reliability Corporation (NERC), *Grid Security Exercise (GridEx II): After-Action Report*, March 2014 and *Grid Security Exercise, GridEx III Report*, March 2016; Scott Heffentrager, PJM Interconnection, "GridEx IV Summary," slide presentation, November 27, 2017, https://www.pjm.com/-/media/committees-groups/committees/mc/20171127-webinar/20171127-item-04-2017-gridex-iv-summary.ashx

[14] Federal Energy Regulatory Commission (hereinafter, FERC), *Reliability Standards for Physical Security Measures*, Order Directing Filing of Standards, Docket No. RD14-6-000, March 7, 2014, p.1, http://www.ferc.gov/CalendarFiles/20140307185442-RD14-6-000.pdf.

necessary because "the current Reliability Standards do not specifically require entities to take steps to reasonably protect against physical security attacks."[15] According to the FERC order, the new reliability standards were to require transmission owners or operators to perform a risk assessment of their systems to identify "critical facilities," evaluate the potential threats and vulnerabilities to those identified facilities, and develop and implement a security plan designed to protect against physical attacks on those identified critical facilities.[16] The order required that each of these steps be verified by NERC or another third party qualified to review them.

On May 23, 2014, NERC filed with FERC its proposal for mandatory physical security standards.[17] On November 20, 2014, FERC approved the proposed standard, with minor changes, as NERC's new *Physical Security Reliability Standard* (CIP-014-1).[18] Following publication in the *Federal Register*, FERC's order approving the standard became effective on January 26, 2015.[19] FERC approved a revised version of the standard (CIP-014-2) on July 14, 2015.[20] Required compliance for the standard began on October 1, 2015 with completion of the final parts required by November 24, 2016 for all applicable entities.

Physical Security Standard Requirements

The stated purpose of NERC's physical security reliability standard is "to identify and protect transmission stations and transmission substations, and their associated primary control centers that if rendered inoperable or

---

[15] FERC, March 7, 2014, p. 2.

[16] FERC, March 7, 2014, pp. 3-4.

[17] NERC, Petition of the North American Electric Reliability Corporation for Approval of Proposed Reliability
Standard CIP-014-1, May 23, 2014, http://www.nerc.com/FilingsOrders/us/NERC%20Filings%20to%20FERC%20DL/Petition%20-%20Physical%20Security %20CIP-014-1.pdf.

[18] FERC, "Physical Security Reliability Standard," Docket No. RM14-15-000, Order No. 802, November 20, 2014.

[19] NERC, "Physical Security Reliability Standard Implementation," January 16, 2015, https://docplayer.net/21487741-Physical-security-reliability-standard-implementation.html.

[20] FERC, letter order to the North American Electric Reliability Corporation, Docket No. RD-15-4-000, July 14, 2015, http://www.nerc.com/FilingsOrders/us/FERCOrdersRules/Letter_Order_CIP-014_20150714_RD15-4.pdf.

damaged as a result of a physical attack could result in instability, uncontrolled separation, or cascading within an interconnection."[21] It applies to transmission owners with assets operating at 500 kV or higher as well as owners with substations operating between 200 kV and 499 kV if they meet certain interconnection or load-carrying criteria.[22] The standard, generally referred to as "CIP-014," consists of six principal requirements (R1-R6), summarized as follows:

R1. Risk assessments by transmission owners to identify critical transmission facilities;
R2. Independent third party verification of risk assessments conducted under R1;
R3. Requirement for transmission owners with critical facilities identified under R1 but not under their operational control to notify the transmission operator of these facilities;[23]
R4. Mandatory threat and vulnerability assessments for critical facilities conducted by transmission owners and operators;
R5. Development, documentation, and implementation of physical security plans to protect critical facilities; and
R6. Independent third party review of the threat and vulnerability assessments performed under R4 and security plans developed under R5.[24]

The standard also lays out a process for compliance monitoring and assessment including audits, self-certifications, spot checking, violation

---

[21] NERC, *CIP-014-2 – Physical Security*, printed December 5, 2017, p. 1, available at https://www.nerc.com/_layouts/15/PrintStandard.aspx?standardnumber=CIP-014-2&title=Physical%20Security&jurisdiction=United%20States. (Hereinafter CIP-014-2). This report uses the terms "critical assets" and "critical substations" to mean "critical transmission stations and transmission substations" as defined under the CIP-014 standard.

[22] CIP-014-2.

[23] A regional transmission operator (RTO) administers the transmission grid for multiple transmission owners in a specified region in accordance with FERC Order No. 2000. RTOs and independent system operators (ISOs) are defined in Section 3 of the Federal Power Act (16 U.S.C. 796).

[24] CIP-014-2, pp. 3-6.

investigations, self-reporting, and handling complaints.[25] The new standard is enforced by NERC or another Regional Entity under a penalty review policy for mandatory reliability standards approved by FERC subject to the Commission's enforcement authority and oversight under the Energy Policy Act of 2005 (P.L. 109-58).[26] Monitoring of compliance with the standard is further discussed below.

FEDERAL OVERSIGHT AND SUPPORT

Three entities play key roles in standards oversight and implementation support for bulk power physical security. NERC and FERC directly oversee implementation of the CIP-014 standards, while the Department of Energy (DOE) plays a supporting role in helping bulk power asset owners to protect their critical assets.

NERC's Implementation Oversight

As stated above, with oversight by FERC, NERC has the authority to develop, oversee, and enforce implementation of the CIP-014 physical security standard.[27] NERC carries out these functions together with the eight Regional Entities (e.g., Midwest Reliability Organization) with which NERC has agreements to delegate its authority to monitor and enforce reliability standards compliance.[28] Collectively, NERC and the Regional Entities comprise the Electric Reliability Organization (ERO) Enterprise.

---

[25] CIP-014-2, p.8.

[26] FERC, *Statement of Administrative Policy on Processing Reliability Notices of Penalty and Order Revising Statement in Order No. 672*, Docket Nos. AD08-6-000 and RM05-30-002, April 17, 2008.

[27] NERC's authorities to monitor compliance with its reliability standards and impose financial penalties are found in FERC regulations at 18 C.F.R. 39.7.

[28] See NERC, "Key Players," web page, March 13, 2018, http://www.nerc.com/AboutNERC/keyplayers/Pages/default.aspx.

In general, NERC employs a risk-based framework to monitor compliance of all its grid reliability standards on the belief that monitoring and enforcement must be "right-sized" based on considerations including risk factors and management practices related to detecting, assessing, mitigating, and reporting of noncompliance.[29]

> As reliability risk is not the same for all registered entities, the Framework examines [bulk power system] risk of registered entities both collectively and individually, to determine the most appropriate [Compliance Monitoring and Enforcement Program] tool to use when monitoring a registered entity's compliance with NERC Reliability Standards. The Framework also promotes an examination into how registered entities operate and tailor compliance monitoring focus to areas that pose the greatest risk to [bulk power system] reliability.[30]

NERC's approach offers flexibility in both the frequency and type of compliance monitoring (e.g., offsite or onsite audits, spot checks, or self-certifications) applied to an entity under a particular standard based on its particular level of reliability risk.[31] To support its compliance approach, NERC may conduct various activities, such as publishing guidance documents, providing training, and conducting outreach, "to promote transparency and confidence" in the utility industry's implementation of a standard.[32]

In monitoring compliance of the CIP-014 standard, NERC's focus in 2015 and 2016 was on the standards' requirements to identify critical transmission stations and substations (Requirements R1 and R2), ensuring that this identification was "appropriate and risk-informed."[33] NERC required covered entities to self-certify with respect to: risk-assessment,

---

[29] NERC, *Overview of the ERO Enterprise's Risk-Based Compliance Monitoring and Enforcement Program*, September 5, 2014, p. iv.

[30] NERC, *2017 ERO Enterprise Compliance Monitoring and Enforcement Implementation Plan, Version 2.5*, May 2017, p. 3.

[31] NERC, May 2017, p. 3.

[32] NERC, "Physical Security Reliability Standard Implementation," January 16, 2015, p. 3, https://docplayer.net/21487741-Physical-security-reliability-standard-implementation.html.

[33] NERC, May 2017, p. 16.

identifying critical assets, and third party verification. NERC also conducted voluntary outreach through on-site visits with 19 covered entities to discuss security measures and CIP-014 implementation challenges.[34] In cases where there have been discrepancies between utility-generated critical asset lists and critical assets identified by the independent third parties, NERC has required the covered entities to provide further information and explanation to address the discrepancy. NERC has also been conducting audits of entities which have identified more, or fewer, critical substations as a percentage of all their substations than is typical.[35] The detailed findings of NERC's compliance activities are not publically disclosed due to the confidential nature of security information. However, NERC stated that, based on observations in 2016, the utility industry was "making progress towards effective implementation of and compliance with CIP-014-2."[36] A NERC presentation about its voluntary and informal site visits reported "remarkable progress" on physical security among 19 asset owners visited as of February 2016.[37]

In 2017, NERC increased its focus on the scope of utility security plans (R5), including their timelines for implementing security measures and the utility industry's overall progress in implementing CIP-014. The ERO Enterprise has prioritized auditing the quality of covered entities' risk management plans. In the second quarter of 2017, compliance audit staff were provided with guidance and training on bulk power physical security best practices as a reference for evaluating the physical security measures implemented by the covered entities.[38]

---

[34] NERC, *2016 ERO Enterprise Compliance Monitoring and Enforcement Program Annual Report*, February 8, 2017, p. 18, http://www.nerc.com/pa/comp/CE/Compliance%20Violation%20Statistics/2016%20Annual%20CMEP%20Report.pdf.

[35] NERC, Staff meeting with CRS analysts, Washington, DC, December 7, 2017.

[36] NERC, May 2017, p. 16.

[37] Carl Herron, NERC, "CIP-014-02 Physical Security Site Visits," slide presentation, April 14, 2016, https://www.frcc.com/Compliance/EducationalMaterials/Educational%20Materials/Workshops%20-%20Workshop%20Event%20Materials/2016-04%20-%20OP%20Spring%20Compliance%20Workshop%20(April%2012-14)/7.%20CIP-014-2%20Physical%20Security%20Site%20Visits.pdf.

[38] NERC, *Compliance Monitoring and Enforcement Program Quarterly Report, Q2 2017*, August 9, 2017, p. 8, http://www.nerc.com/gov/bot/BOTCC/Compliance%20Committee%202013/Compliance%20Committee%20Open%20Meeting%20-%20August%209%202017.pdf.

The ERO Enterprise expects to complete audits of the largest entities within three years of the effective date of CIP-014. As of February 2018, NERC had conducted compliance audits of approximately 45% of the covered entities with critical transmission stations and substations as defined under CIP-014. NERC had also audited over 30% of entities that did not identify critical assets after applying the CIP-014 criteria (under R1). NERC staff expects to have audited approximately 70% of the entities with CIP-014 critical assets by the end of 2018.[39] According to its stated schedule, NERC would audit the remaining entities in 2019. Subsequent monitoring and enforcement will focus more heavily on implementation of measures in the grid security plans.

According to NERC, the audits completed to date have not uncovered any major compliance failures, and NERC has been "encouraged" by security measures that utilities have put in place so far.[40] NERC has found no serious risk violations of the CIP-014 standard. Of 19 noncompliance issues identified, 8 were found to be "minimal" or "moderate" risk, with 2 warranting a financial penalty. The remaining 11 noncompliance issues are under review.[41]

Electricity Information Sharing and Analysis Center

In addition to its standards activities, NERC also supports security of the electric power sector as the operator of the Electricity Information Sharing and Analysis Center (E-ISAC). Established in 1998, the E-ISAC is the electricity sector's primary communications channel for security-related information, situational awareness, incident management, and coordination.[42] Among its key responsibilities, the E-ISAC gathers and analyzes security data, shares it with stakeholders, and communicates

[39] NERC, email to CRS, February 14, 2018.
[40] NERC, December 7, 2017.
[41] NERC, February 14, 2018.
[42] ISACs for critical infrastructure sectors were established under Presidential Decision Directive 63, May 22, 1998. NERC operates the E-ISAC in collaboration with the Department of Energy and the Electricity Subsector Coordinating Council (ESCC). The ESCC, established in 2004 by companies in the electric power industry, coordinates policy-related activities involving the reliability and resilience of the sector, including physical and cyber infrastructure.

security risk mitigation strategies.[43] Bulk power entities are required to report physical security events to the E-ISAC under NERC's Event Reporting Reliability Standard (EOP-004), which was approved by FERC in 2013 and revised in 2015.[44]

Although operated by NERC, the E-ISAC is independent and organizationally separate from NERC's standards enforcement functions; information shared by utilities with the E-ISAC is not passed on to NERC compliance staff.[45] Nonetheless, the E-ISAC has played a role in facilitating industry understanding of physical security best practices. For example, the E-ISAC has added significant physical security threats and tactics to the NERC's biennial GridEx security exercises (discussed later in this chapter). In 2015, the E-ISAC also established a Physical Security Advisory Group, which includes industry physical security professionals, outside experts, and representatives from DOE and the Department of Homeland Security (DHS), to assist in the analysis of physical security threats and advise asset owners on physical threat mitigation. Through these efforts, the E-ISAC developed and ratified a design basis threat for the electric sector in December 2015.[46] The E-ISAC also has hosted two threat workshops, with plans for more.[47] Thus, while the E-ISAC has had no role in enforcing the CIP-014 standards, the security risk and mitigation information it develops and promulgates support the activities of bulk power asset owners complying with the standards.

FERC Oversight

As the agency with general statutory authority over grid reliability, and the agency which ordered and approved NERC's CIP-014 standard, the Federal Energy Regulatory Commission also oversees implementation of

[43] NERC, *Understanding Your E-ISAC*, June 2016, p. 3.

[44] NERC, "EOP-004-3—Event Reporting," 2015, http://www.nerc.com/pa/Stand/Reliability%20Standards/EOP-004- 3.pdf.

[45] NERC, June 2016, p. 3.

[46] NERC, *State of Reliability 2016*, May 2016, p. 7.

[47] NERC, *State of Reliability 2017*, June 2017, p. 62.

the standard. In carrying out this oversight, FERC relies primarily on annual compliance reporting by NERC.[48] However the commission also conducts some independent compliance activities, and it also conducts some compliance activities in cooperation with NERC. For example, during the initial rollout of the CIP-014 standard in 2016, FERC staff coordinated with NERC staff in support of on-site visits to the covered entities discussed above.[49]

In its order approving CIP-014-01, the commission stated that NERC staff would submit to both the NERC Board of Trustees and FERC a report following implementation of requirements R1, R2, and R3 about the scope, number, and characteristics of facilities identified as critical.[50] The order stated that

> Based on the results reported by NERC, we expect Commission staff to audit a representative number of applicable entities to ensure compliance with Reliability Standard CIP-014-1. Depending on the audit findings, the Commission will determine if there is a need for any further action by the Commission including, but not limited to, directing NERC to develop modifications to Reliability Standard CIP-014-1 to provide greater specificity to the methodology for determining critical facilities.[51]

As of November 2, 2017, FERC had completed two audits of critical assets identified by covered entities (R1) and was in the process of conducting a third. These audits have involved technical review of utility regulatory documents by FERC engineers. According to FERC staff, the initial audits identified one issue of concern related to the interpretation of specific language in the standard regarding asset criticality.[52] In addition to

---

[48] FERC, *Order on Electric Reliability Organization Reliability Assurance Initiative and Requiring Compliance Filing*, Docket No. RR15-2-000, p. 11, February 19, 2015, http://www.nerc.com/pa/comp/Reliability%20Assurance%20Initiative/FERC_Order_Approving_Risk-Based_CMEP.pdf.

[49] NERC, May 2017, p. 16.

[50] FERC, *Physical Security Reliability Standard*, Docket No. RM14-15-000, Order No. 802, November 20, 2014, p. 23, http://www.nerc.com/FilingsOrders/us/FERCOrdersRules/Final%20Rule%20=on% =20CIP-014-1.pdf.

[51] FERC, Order No. 802, p. 24.

[52] FERC, Staff meeting with CRS analysts, Washington, D.C., November 2, 2017.

NERC's annual reports, FERC receives from NERC periodic Notices of Penalty (NOP) to regulated entities for reliability standards violations. As of November 30, 2017, FERC received NOPs for two violations (apparently at the same utility) of the CIP-014 standard.[53]

DOE Initiatives

Presidential Decision Directive 63 (PDD-63), issued during the Clinton Administration in 1998, established national policy for critical infrastructure protection from both physical and cyber threats.[54] PDD-63 established 15 critical infrastructure sectors. The Department of Energy was assigned responsibility for (1) the electric power, and (2) the oil and natural gas production and storage sectors. The George W. Bush Administration built on the work of PDD-63, superseding it in 2003 with Homeland Security Presidential Directive 7 (HSPD-7) on "Critical Infrastructure Identification, Prioritization, and Protection."[55] HSPD-7 again assigned to DOE (as a Sector-Specific Agency) responsibility for the energy sector—including electric power—as well as responsibility for being the federal coordinator for all critical infrastructure protection efforts.[56] The Obama Administration superseded HSPD-7 with Presidential Policy Directive 21 (PPD-21) on "Critical Infrastructure Security and Resilience" in 2013.[57] PPD-21 retained the Sector-Specific Agencies (SSAs) from HSPD-7, with DOE continuing as the SSA for the energy

---

[53] NERC, *Enforcement and Mitigation*, "Searchable NOP Spreadsheet," web page, accessed December 12, 2017, http://www.nerc.com/pa/comp/CE/Pages/Enforcement-and-Mitigation.aspx.

[54] National Security Council and National Security Council Records Management Office, "PDD-63—Critical Infrastructure Protection," *Clinton Digital Library*, May 20, 1998.

[55] George W. Bush White House Archives, "Critical Infrastructure Identification, Prioritization, and Protection," Homeland Security Presidential Directive/HSPD-7, December 17, 2003.

[56] For details about the roles of Sector-Specific Agencies, see Department of Homeland Security, "Sector-Specific Agencies," web page, July 11, 2017, https://www.dhs.gov/sector-specific-agencies.

[57] Barack H. Obama White House Archives, "Critical Infrastructure Security and Resilience," *Presidential Policy Directive*-21, February 12, 2013.

sector. Thus, DOE has had a supportive role in helping utilities to protect bulk power critical assets over the last two decades.

Until recently, DOE's power grid security activities were led by its Office of Electricity Delivery and Energy Reliability (OE) within the Office of the Under Secretary for Science and Energy. A 2008 OE report stated that "OE's mission is to advance technology—in partnership with industry, government, academia, and the public—to meet America's need for a reliable, efficient, and resilient electric power grid."[58] Although the office was primarily focused on grid cybersecurity, it did conduct activities related to power grid physical security, including analysis of large power transformer security, a substation security awareness campaign, and efforts to support and coordinate research and development for physical security.[59] On February 14, 2018, DOE announced that the Secretary of Energy was establishing a new Office of Cybersecurity, Energy Security, and Emergency Response (CESER) to be led by an Assistant Secretary with responsibilities to help protect energy infrastructure from "from cyber threats, physical attack and natural disaster."[60] How this reorganization will affect DOE's activities in bulk power physical security remains to be seen.

OBSERVED CHANGES IN BULK POWER PHYSICAL SECURITY

Most grid security analysts consider the 2013 Metcalf substation attack to have been the "wake up call" which both changed electric sector attitudes toward grid physical security and motivated the promulgation of NERC's physical security regulations. Since that time, there have been a number of apparent changes within the electricity sector related to

[58] Department of Energy, Office of Electricity Delivery and Energy Reliability (Hereinafter OE), *National SCADA Test Bed Program, Multi-Year Plan FY2008-2013*, January 2008, p. 7.

[59] Department of Energy, Energy Sector-Specific Plan, 2015, pp. 16, 27. For discussion of OE's cybersecurity activities, see CRS Report R44939, *Cybersecurity for Energy Delivery Systems: DOE Programs*, by Paul W. Parfomak, Chris Jaikaran, and Richard J. Campbell.

[60] U.S. Department of Energy, "Secretary of Energy Rick Perry Forms New Office of Cybersecurity, Energy Security, and Emergency Response," press release, February 14, 2018.

increasing bulk power physical security. It is not clear whether these changes have been driven more by changes in utility perceptions of grid threats or by NERC's mandatory security standards. Furthermore, there is currently no comprehensive accounting of changes in physical security throughout the sector. Nonetheless, anecdotal information in the public domain suggests that such changes may be significant and widespread. They are discussed in the following sections.

Corporate Structure Supporting Physical Security

One criticism that arose in the wake of the Metcalf attack was that physical security management at Pacific Gas and Electric Company (PG&E, the Metcalf substation's owner) and at other utilities was not a centrally organized or well-supported function in corporate management. This lack of support limited the influence of security managers in corporate planning and financial decisions.[61] However, it appears that many utilities have been reconfiguring and elevating physical security functions within their corporate structures. For example, owners of transmission assets such as PG&E, American Electric Power, and Xcel Energy have appointed Chief Security Officers at senior levels responsible for managing both physical and cyber security risks company-wide.[62]

> The senior security professional, typically at the vice president or director level, now has direct access to the [Chief Executive Officer] and company boards of trustees, often to supply situational awareness of

[61] See, for example: Tony Kovaleski, Liz Wagner, and Mark Villarreal, "Internal Memo Reveals PG&E Years Away from Substation Security," *NBC Bay Area*, April 5, 2106, https://www.nbcbayarea.com/investigations/Internal-Memo-Reveals-PGE-Years-Away-from-Substation-Security-303833811.html

[62] PG&E Corp., "Bernard A. Cowens," web page, January 9, 2017, http://www.pgecorp.com/corp/about-us/officers/company/bernard-cowens.page; American Electric Power, "AEP Names Partlow Vice President & Chief Security Officer," press release, August 25, 2015; Xcel Energy, *Application of Southwestern Public Service Company for Authority to Change Rates*, Direct Testimony of Stephen J. Brown, filing with the Public Utility Commission of Texas, August 21, 2017, https://www.xcelenergy.com/staticfiles/xe-responsive/Company/Rates%20&%20Regulations/Rate%20Cases/Brown-RR-Direct.pdf.

> physical and cybersecurity issues.... The electricity industry is quickly moving away from security as an "addition duty".... [M]ost utilities today have dedicated security departments committed to the protection of company assets and personnel.[63]

Utilities are also centralizing and bolstering their physical security capabilities at the operational level. Between 2014 and 2017, for example, Xcel Energy consolidated and grew its staffing for the "Chief Security Officer class of services" from 47 to 63 employees.[64] According to the company's regulatory filings,

> the increase in average staffing levels.... was due to the need to correct a lack of resources to ensure adequate headcount to provide essential cyber and physical Enterprise Security services for Xcel Energy.... This increase in staffing demonstrates the emerging need that led to a stand-alone organization (i.e., the Chief Security Officer) to focus on Cyber Operations, Enterprise Resilience, Physical Security and Security Governance.[65]

Likewise, in response to the Metcalf attack, Dominion Energy established "a true cross-functional team with more than 100 people representing the entire Dominion organization," to develop and implement a more comprehensive substation security program.[66] Such efforts appear to extend to major publicly owned utilities as well. For example, according to the head of the Western Area Power Administration (WAPA), one of four federal power marketing administrations,

---

[63] Brian Harrell, "The Modern Look of a Utility's Chief Security Officer," *CSO*, August 4, 2016, https://www.csoonline.com/article/3101474/leadership-management/the-modern-look-of-a-utilitys-chief-security-officer.html

[64] Xcel Energy, Application of Southwestern Public Service Company for Authority to Change Rates, Update Testimony of Stephen J. Brown, September 27, 2017, p. 10, https://www.xcelenergy.com/staticfiles/xe-responsive/Company/Rates%20&%20Regulations/Rate%20Cases/13%20-%20BrownRRUpdate.pdf

[65] Xcel Energy, August 21, 2017, p. 26.

[66] Bob McGuire, et al., "Substation Security Is More Than Just a Fence," *T&D World*, September 28, 2015.

> WAPA's approach to physical security.... began in 2013 with the consolidation of our Office of Security and Emergency Management across our five regions and the implementation of a sophisticated risk-based program in analyzing the threats and vulnerabilities to our substations.[67]

The Tennessee Valley Authority (TVA), which operates federally-owned hydroelectric and nuclear generation and associated transmission assets, recently closed a job posting for eight entry-level Inspectors, each to be "trained as a physical security specialist" to provide "comprehensive security services, including assessments of facilities to identify credible threats, and implementation and testing of countermeasures to mitigate risks."[68]

Some transmission owners are also specifically increasing their in-house intelligence capabilities in physical security, including recent postings for positions such as "Security Intelligence Specialist" and "Director—Corp Security Info & Intelligence."[69] While the examples above are anecdotal, they would be consistent with what may be a trend among key grid owners to make physical security a better-organized and more influential corporate function. Not all utilities may be implementing such organizational changes, however.

Physical Security in Long-Term Transmission Planning

Since NERC promulgated the CIP-014 standards, some utilities have begun to put a greater emphasis on bulk power physical security as a

---

[67] Mark A. Gabriel, Administrator and Chief Executive Officer, Western Area Power Administration, "Physical and Cyber Threats," *T&D World*, May 8, 2017. Power Marketing Administrations (PMAs) operate electric transmission systems and sell power generated by federally-owned hydroelectric dams across much of the United States.

[68] Tennessee Valley Authority, "Inspector I – 507038," job posting, *Linkedin JOBS*, web page, posted January 17, 2018, accessed February 1, 2018.

[69] American Transmission Company, "Security Intelligence Specialist," job listing on *LinkedIn*, posted March 6, 2017; Avangrid, "Director—Corp Security Info & Intelligence," job listing on Glassdoor.com, posted January 3, 2018.

design consideration in long-term transmission system planning. This approach aligns with the California Public Utilities Commission's recommendation in its 2018 report that, "there should be an emphasis on incorporating a menu of physical security strategies [into] any substation from the time of its inception."[70] For example, Public Service Enterprise Group's transmission planning criteria for its Long Island system in New York discusses the use of power system simulation tools for "various transmission system security and reliability studies."[71] Commonwealth Edison's transmission planning criteria includes a separate section on "security criteria" for system design which considers "severe low probability outage combinations" and seeks "to avoid cascading outages, instability, or widespread blackout."[72] Such criteria could apply to both natural and man-made outages, but they are consistent with, and readily applied to, design considerations for enhanced physical security. American Electric Power (AEP) also has incorporated asset criticality as a design criterion in its transmission planning.

> As a result of the revised NERC CIP standards, AEP now classifies all of its bulk electric system facilities based on the critical nature of the equipment to determine the level of security needed. This approach allows us to design security controls directly into new infrastructure from the start, building the costs into capital projects as needed. It also allows us to be more proactive with new and existing infrastructure while balancing risks with mitigation solutions.[73]

In its plans for a 2018 reliability-related upgrade at one its substations, Vermont Electric Power Company states that it "will also take the opportunity to make improvements to the physical security" of the

---

[70] CPUC, January 2018, p. 8.

[71] PSEG Long Island, "Transmission Planning Criteria," accessed January 10, 2018, p. 5, https://www.psegliny.com/files.cfm/TransmissionPlanningCriteria.pdf.

[72] Commonwealth Edison Co., "Transmission Planning Criteria," February 10, 2017, p. 10, https://www.pjm.com/-/media/planning/planning-criteria/commonwealth-edison-planning-criteria.ashx?la=en

[73] American Electric Power Corp., *2017 AEP Corporate Accountability Report*, "Cyber and Physical Security," web page, May 25, 2017, http://www.aepsustainability.com/about/security/cyber.aspx

substation.[74] According to NERC officials, based on security criteria, some utilities also have begun to consider new transmission interconnections not only to increase line capacity for bulk power flows, but also to reduce the criticality of particular transformer substations in congested areas by providing more transmission paths around them.[75]

New Security Products and Services

As utilities have devoted greater organizational and financial resources towards power grid physical security, industry vendors have been offering more physical security products and services to meet sector demand. As one utility services company has observed, "we can expect plenty of innovation as manufacturers see new markets due to the new standards for physical security of critical substations."[76] These offerings range from analytical services for security planning to physical products to harden physical assets. A comprehensive survey of such offerings is beyond the scope of this chapter, but the following examples illustrate the kinds of products now commercially available in the bulk power physical security market.

- *Security Program Planning and Implementation.* Engineering and security consulting firms have developed customizable programs specifically for power grid physical security review, planning, analysis, and implementation in compliance with the CIP-014 standards and utility-specific requirements.[77]
- *Anti-Intrusion Products.* Vendors have been marketing existing intrusion-related products specifically for use at bulk power critical

---

[74] Vermont Electric Power Company, "East Avenue & Queen City Substation Improvement Project," web page, accessed February 1, 2018, https://www.velco.com/our-work/projects/project-east-avenue-queen-city-substation-improvement-project

[75] NERC, December 7, 2017.

[76] Southwire Company, "Protecting the Grid," *T&D World*, sponsored content, May 15, 2017.

[77] See, for example: Burns & McDonnell, "Station Defender," web page, January 30, 2018, https://info.burnsmcd.com/station-defender/project-delivery; Corporate Risk Solutions, "Physical Security," web page, January 30, 2018.

facilities. These products include visual, acoustic, thermal radar, and electromagnetic systems for facility monitoring, intrusion detection, and response.[78]

- *Hardened Transformers and Components.* At least two major manufacturers have been marketing bulk power transformers with integrated ballistic shielding, or customizable plates to shield existing transformers.[79] Smaller manufacturers have also begun marketing hardened transformer components, such as composite bushings, for new and retrofit substation applications.[80]
- *Substation Perimeter Shielding.* A number of vendors have been marketing perimeter fencing and wall products specifically for visual and physical shielding of bulk power substations.[81] Most of these products are designed specifically to protect against rifle attacks such as the Metcalf attack.

Although new physical security products and services are being marketed in the utility sector, there is no comprehensive source of data about their sales to bulk power asset owners. Simply because vendors are marketing products does not mean that many utilities are buying them. For example, as of October 2017, Siemens Corp. had announced only one commercial order for its new transformer ballistic shielding retrofit

---

[78] See, for example: "How VTI Security Protected an Electrical Substation With a Radar-Thermal Imaging Solution," Security *Sales & Integration*, September 20, 2017, https://www.securitysales.com/in-depth/vti-security-radar-thermal-imaging-solution/; and i2c Technologies, Ltd., "Power Substation Protection," marketing brochure, May 2017, http://www.i2ctech.com/wp-content/uploads/2017/05/2509-i2cTech-CMYK.pdf.

[79] See, for example: Siemens AG, "First Bullet Resistant Retrofit Ordered for a Transformer," press release, accesses January 28, 2018, https://www.siemens.com/global/en/home/products/energy/references/first-bullet-resistant-retrofit-ordered-for-a-transformer.html

[80] Mike Sheppard and Saqib Saeed, "Bullet and Weather Concerns Driver of Retrofits in US Market," Power Technology Research LLC, October 26, 2017, https://powertechresearch.com/bullet-and-weather-concerns-driver-ofretrofits-in-us-market/.

[81] See for example: Oldcastle, Inc., "How Precast Substation Walls Increase Power Grid Security," web page, https://www.buildingsolutions.com/industry-insights/how-precast-substation-walls-increase-power-grid-security; AFTEC LLC, "Substation Security Walls," web page, 2017, https://aftec.com/substation-security-walls/;

product.[82] Thus, the overall impact of such offerings on the sector cannot be qualified reliably. Additional discussion of physical security spending is in the following section.

Capital Investment in Physical Security

Major changes in power grid operational expenses and capital investment are generally slow to occur. In privately owned utilities, significant changes in spending and plans for new capital projects may need to go through a number of rigorous screens, including power network modeling, a corporate capital allocation process, a regulatory approval process, and a procurement process. Publicly owned utilities may need approval from cooperative boards, or municipal or federal officials. This combination of requirements can take years to complete. Consequently, many significant operating expenditures or capital investments for physical security identified in security plans under CIP-014 may still be working their way through utility budgets and implementation. For example, in a 2016 rate filing, Southern California Edison stated that it planned to make physical security improvements at approximately 24 facilities in 2015-2017 and proposed to upgrade 8 substations per year from 2016 through 2020.[83] Likewise, in its 2016 annual report, Dominion Resources' timeline for power grid capital investment in "Physical Security" runs to 2021.[84]

Notwithstanding the potential length of time it may take for some security projects to be approved and implemented, there are indications in the public record that bulk power asset owners have already been spending more on new physical security measures. In its December 2016 report, the

---

[82] Siemens AG, "First Bullet Resistant Retrofit Ordered for a Transformer," press release, October 17, 2017, https://www.siemens.com/global/en/home/products/energy/references/first-bullet-resistant-retrofit-ordered-for-a-transformer.html

[83] Southern California Edison Co., Application Of Southern California Edison Company (U 338E) For Authority To Increase Its Authorized Revenues For Electric Service In 2018, Among Other Things, And To Reflect That Increase In Rates, A.16-09-001, Before the Public Utilities Commission of the State of California, September 1, 2016, http://www3.sce.com/sscc/law/dis/dbattach5e.nsf/0/9F664E3F0B77B7E488258195007C8F53/$FILE/SCE%20Opening%20Brief%20and%20COS.pdf.

[84] Dominion Resources, Inc., *Energy is Essential*, 2016 Summary Annual Report, 2017, p. 5.

Edison Electric Institute stated that "primary factors driving transmission investment between 2015 and 2019" included "system hardening and resiliency to minimize adverse catastrophic events" and "improvements to comply with evolving transmission reliability and security compliance standards."[85] In its January 2018 white paper, the California Public Utilities Commission (CPUC) reports that investor-owned utilities under its jurisdiction "already ... have sought approval for tens of millions of dollars in General Rate Case funding to ensure physical security."[86] The following examples illustrate the types of physical security projects and recent spending in publicly available sources.

- In 2017, the Bonneville Power Administration announced stand–alone plans to install security fencing at two high-voltage substations in compliance with NERC's security standards and to "protect critical assets from theft, vandalism, and terrorism."[87]
- In 2017, PPL Electric Utilities reportedly filed for regulatory approval for a $450,000 expenditure to reconfigure a 500 kV substation in compliance with NERC's CIP-014 physical security standard.[88]
- In 2017 regulatory filings, Vectren (Indiana) described plans to invest $2.9 million for physical security upgrades at critical substations, including enhanced fencing, access control, video surveillance, and perimeter motion detection.[89]

---

[85] Edison Electric Institute, *Transmission Projects: At A Glance*, December 2016, p. vi.

[86] California Public Utilities Commission (CPUC), *Security and Resilience for California Electric Distribution Infrastructure: Regulatory and Industry Response to SB 699*, January 2018, p. 5.

[87] Bonneville Power Administration, Categorical Exclusion Determination, "Proposed Action: Covington and Maple Valley Substations Perimeter Security Upgrades," April 27, 2017, https://www.bpa.gov/efw/Analysis/CategoricalExclusions/cx/20170427_Covington-and-Maple-Valley-Substations-Perimeter-Security-Upgrades.pdf.

[88] Corina Rivera Linares, "PPL Electric Utilities Seeks Approval of Two Projects in Pennsylvania," *Transmission Hub*, PennWell Publishing, May 22, 2017.

[89] Southern Indiana Gas and Electric Company d/b/a Vectren Energy Delivery of Indiana, Inc. IURC Cause No. 44910, filing with the Indiana Utility Regulatory Commission, February 23, 2017, Attachment LKW-2, p. 31, https://iurc.portal.in.gov/legal-case-details/?id=3b675b4f-eff9-e611-80fd-1458d04e2f50

- According to the Western Area Power Administration, its expenses for physical security "nearly tripled" between 2013 and 2017.[90]

Utility Participation in Voluntary Security Programs

Although the CIP-014 mandatory physical security standards have only been in effect since 2014, bulk power asset owners have had earlier opportunities to participate in voluntary security initiatives administered by NERC and DHS. Utility participation in these voluntary programs is another indication of overall efforts in the sector to improve critical asset physical security.

NERC Grid Security Exercises

In 2011, NERC conducted GridEx, the first of an ongoing series of biennial electric sector-wide grid security exercises.[91] The 2011 exercise assessed the readiness of utilities to respond to a cyberattack, strengthened their crisis response, and provided input for internal security program improvements. Although the exercise was focused on a cyberattack, it did involve physical incursions into power grid substations as well as aspects of grid monitoring and recovery that would be relevant to an attack on critical transformers.[92] After the Metcalf attack in 2013, NERC conducted a second, more expansive grid security exercise, GridEx II. The exercise scenario included a cyberattack on the grid coupled with a coordinated physical attack against a subset of transmission and generation assets—including critical transformer substations.[93] NERC conducted GridEx III in 2015, again including a baseline scenario with cyber and physical attacks, but also with an option for participants to customize the baseline scenario

---

[90] Mark A. Gabriel, May 8, 2017.

[91] NERC's E-ISAC division organizes and administers its GridEx exercises.

[92] North American Electric Reliability Corporation (NERC), *2011 NERC Grid Security Exercise: After Action Report*, March 2012, p. i.

[93] NERC, *Grid Security Exercise (GridEx II): After-Action Report*, March 2014, p.15; Matthew L. Wald, "Attack Ravages Power Grid. (Just a Test.)," *New York Times*, November 14, 2013.

to meet local objectives.[94] NERC conducted its most recent exercises, GridEx IV, in November 2017.

According to NERC, one indication of progress in bulk power grid security is increasing participation by electricity sector entities in its GridEx exercises. The number of utilities participating in GridEx rose from 49 in 2011 to 166 in 2015.[95] NERC has not yet released participation details for GridEx IV, but the DOE reported that the latest exercise had more participants than in 2015.[96]

DHS Critical Infrastructure Surveys

The Department of Homeland Security's Protective Security Coordination Division conducts voluntary field assessments of critical infrastructure to identify vulnerabilities, interdependencies, capabilities, and cascading effects of potential terrorist attacks. As part of these efforts, DHS Protective Security Advisors offer voluntary, web-based security surveys of critical facility security using the agency's Infrastructure Survey Tool developed in 2008. The key goals of the surveys are to identify facilities' physical security and security management, identify security gaps, create facility protective and resilience measures indices that can be compared to similar facilities, and track progress toward improving security.[97] According to DHS officials, of more than 6,000 surveys completed since the program began, over 600 have been conducted on electric power facilities—although the timing of these surveys and the specific types of power facilities involved are not reported.[98]

---

[94] NERC, *Grid Security Exercise: GridEx III Report*, March 2016, p. 7.

[95] NERC, March 2016, p. 1.

[96] U.S. Department of Energy, "GridEx IV: Government and Industry Exercise Together to Improve the Response to Grid Security Emergencies," November 21, 2017, https://www.energy.gov/articles/gridex-iv-government-and-industry-exercise-together-improve-response-grid-security

[97] Department of Homeland Security, "Critical Infrastructure Vulnerability Assessments," web page, April 17, 2017, https://www.dhs.gov/critical-infrastructure-vulnerability-assessments

[98] Daniel Genua, Department of Homeland Security, Presentation at George Mason University, Center for Energy Science and Policy, Grid Security Symposium, Arlington, VA, October 25, 2017, http://cesp.gmu.edu/wp-content/uploads/2017/10/UNCLASS_GMU-Panel-Presentation_25Oct2017_FINAL.pdf

LEGISLATIVE PROPOSALS IN THE 115TH CONGRESS

Given the relatively recent promulgation of NERC's new physical security standards, bulk power physical security has not been a major legislative focus in the 115th Congress. Nonetheless, several bills include provisions intended to enhance bulk power physical security—primarily by establishing new DOE grid security programs rather than by imposing new requirements on FERC or on bulk power asset owners directly. The relevant provisions of these bills, and a related resolution, are summarized below.

- The *Enhancing Grid Security Through Public-Private Partnerships Act* (H.R. 5240) would require DOE to establish a program to facilitate public-private partnerships for electric utility physical security and cybersecurity, among other provisions. Program activities would support voluntary implementation of maturity models, self-assessment, and security auditing; sharing of best practices and data collection in the electric sector; and training and technical assistance to utilities (§2(a)).
- The *Energy Emergency Leadership Act* (H.R. 5174) would amend the Department of Energy Organization Act to include "energy emergency and energy security" to the functions assigned to Assistant Secretaries. These functions would include responsibilities with respect to emerging threats, supply, and emergency planning, among others. They would also include "provision of technical assistance, support, and response capabilities with respect to energy security threats, risks, and incidents" (§2).
- The *Energy and Natural Resources Act of 2017* (S. 1460) would require DOE to develop an advanced energy security program to secure energy networks, including electric transmission and delivery. Eligible activities would include developing "capabilities to identify vulnerabilities and critical components that pose major risks to grid security if destroyed or impaired," modeling national

level impacts from human-made events, developing a physical security maturity model, conducting grid security exercises, conducting research on critical asset hardening, and other related measures (§2002(e)).

- The *Leading Infrastructure for Tomorrow's America Act* (H.R. 2479) would establish a grant program administered by DOE "to enhance energy security through measures for electricity delivery infrastructure hardening and enhanced resilience and reliability" (§31101(a)).
- The *Advancing Grid Storage Act of 2017* (S. 1851) would establish a competitive grant program for pilot energy storage systems administered by DOE with one objective being to "improve the security of critical infrastructure and emergency response systems" in the electric grid (§5(a)(4)(A)).
- The *Grid Cybersecurity Research and Development Act* (H.R. 4120) would require DOE, together with bulk power asset owners, and in collaboration with the National Laboratories, to "utilize a range of methods, including voluntary vulnerability testing and red team-blue team exercises, to identify vulnerabilities in physical and cyber systems" (§6(a)).
- The *Flexible Grid Infrastructure Act of 2017* (S. 1875) would require DOE to: develop model standards for the electric distribution grid, in part to improve security with respect to physical threats (§5(d)(1)), evaluate whether new performance standards and testing procedures are needed to ensure electrical equipment resilience in the face physical threats (§5(d)(2)), and submit to Congress methods and guidelines for calculating the costs and benefits of investments in resilience and security solutions for the electric grid (§5(e)(1)).
- *House Resolution 334* states that it should be the policy of the United States to, among other things, "bolster the reliability, affordability, diversity, efficiency, security, and resiliency of domestic energy supplies, through advanced grid technologies,"

and to promote advanced grid tools "to increase data security, physical security, and cybersecurity awareness and protection."

POLICY ISSUES FOR CONGRESS

Although NERC's CIP-014 standards have been promulgated, and bulk power asset owners have begun enhancing physical security, Congress continues to be concerned about the current state of electric grid physical security. Among many issues of potential interest, Congress may focus on several with overarching policy significance: security implementation oversight, cost recovery, hardening vs. resilience, and the quality of threat information.

Oversight of Physical Security Implementation

Although FERC's statutory authority for grid reliability and NERC's reliability standards both include provisions for oversight and enforcement, congressional oversight of physical security implementation may be a challenge for several reasons. First and foremost, information about physical security measures is inherently sensitive and there are both statutory and regulatory restrictions on its disclosure.[99] Therefore, the level of security-related information that utilities are willing or able to provide outside the CIP-014 third-party review process or NERC compliance audits is more limited than reports about, say, general reliability or safety.

NERC is not compiling a centralized database of critical assets or security measures implemented by the utilities subject to its physical security standard. Moreover, while NERC may provide security information to FERC, the security-related information NERC can provide in public reports is limited and typically redacted. Therefore, although information about CIP-014 implementation exists among the utilities and

[99] FERC regulations for the submission, designation, handling, sharing, and dissemination Critical Energy/Electric Infrastructure Information (CEII) are at 18 C.F.R. § 388.113.

independent third parties (operating within the standard), and is provided at some level of specificity to NERC, that information may not be as useful or visible as it could be to Congress or other outside entities.

Another oversight challenge arises because NERC's CIP-014 standards are not prescriptive; bulk power asset owners have considerable discretion in the nature and timing of the physical security measures they may include in their physical security plans. NERC viewed such flexibility as necessary for its standard due to the unique characteristics of each utility's bulk power system and the risks it faces. However, this flexibility also may make it more difficult to develop useful metrics for CIP-014 implementation and comparing implementation among asset owners. NERC's standards for power grid physical security may ensure considerable consistency in the *process* utilities must undertake to identify critical substations and develop plans to secure them. However, they may not ensure consistency among the various security plans nor in the specific measures the individual asset owners will choose to implement to reduce the risk of intentional attacks. For example, ballistic shielding at critical substations may be an appropriate and sufficient protective measure for some utility assets, say, in open and rural areas, but not necessarily in more urban areas.

Even when detailed company-specific information about physical security measures is available, it might be difficult to develop reliable metrics to evaluate it. Metrics are an important tool NERC uses to evaluate utility performance in the context of power grid reliability.[100] However, officials at EEI have stated that measuring the adequacy of grid security for a diverse set of asset owners under changing risk circumstances poses significant problems. "Security metrics (for both cyber and physical security) have consistently been a challenge due evolving threats and vulnerabilities. If you build an eight-foot fence, the attacker just needs to

[100] See NERC, "Reliability Indicators," web page, http://www.nerc.com/pa/RAPA/Pages/ReliabilityIndicators.aspx.

bring a nine-foot ladder."[101] NERC is actively engaged in efforts to develop bulk power system security metrics in which it has likewise encountered "challenges associated with developing relevant and useful security metrics that rely on data willingly and ably provided by individual entities."[102]

Congress may judge the effectiveness of the CIP-014 physical security standards as best it can based on reports and testimony from NERC and FERC as well as information from the assets owners themselves. However, due to the issues above, if Congress decides the information as currently structured is insufficient to draw reliable conclusions about the status of bulk power physical security as a whole, it may revisit how the responsible agencies collect, measure, and report it. Congress may also consider additional avenues for reviewing this information, for example, through classified briefings or specifically requested studies or reports. Also, as FERC continues to implement its policy of regulating physical security of the power grid, Congress may examine whether company-specific security initiatives appropriately reflect the risk profiles of their particular assets, and whether additional security measures across the grid overall uniformly reflect terrorism risk from a national perspective.

Financial Requirements and Cost Recovery

Two of the barriers to physical security investment among utilities prior to the Metcalf attack were competition for limited capital investment resources and justifying security spending to corporate boards and utility rate regulators. NERC regulatory requirements for physical security make it easier for security managers to justify related operating and capital expenditures to corporate leadership, and to seek cost recovery for such expenditures through regulated rates. However, even where regulators have been supportive of cost recovery for physical security investments in general, they have faced challenges gauging the prudence of specific

---

[101] Chris Hickling, Edison Electric Institute, "RE: CIP-014 Implementation Update," email to CRS, October 30, 2017.

[102] NERC, *State of Reliability 2017*, June 2017 p. vii. For an expansive discussion of NERC's efforts to develop security metrics, see Appendix G in this NERC report.

security investments because they are hard to evaluate on a traditional benefit-cost basis. As a 2006 report from the Electric Power Research Institute states,

> Security measures, in themselves, are cost items, with no direct monetary return. The benefits are in the avoided costs of potential attacks whose probability is generally not known. This makes cost-justification very difficult.[103]

Note that cost-justification requires not only the approval of utility management, but also of FERC and potentially state public utility commissions which regulate the rates grid owners may charge for electric transmission and distribution service. Regulators are responsible for ensuring that electricity rates are just and reasonable. They must be convinced that any new grid security capital costs and expenses are necessary and prudent before they will allow them to be passed through to ratepayers. However, corporate financial processes differ from utility to utility, and utility rate regulation differs from jurisdiction to jurisdiction, so investment and cost recovery for physical security is not uniform across the electricity sector and remains a work in progress. As implementation of new physical security plans under CIP-014 continues, Congress may examine whether the overall level of investment appropriately reflects the level of security risk facing the bulk power system, and whether any cost-recovery barriers are preventing assets owners from making investments necessary to secure the grid.

Hardening vs. Resilience

There are two fundamental approaches to reducing the risk of a successful physical attack on the electric grid. The first approach, which is

---

[103] Electric Power Research Institute (EPRI), *Technologies for Remote Monitoring of Substation Assets: Physical Security*, March 2006, p. viii.

the principal approach of NERC's CIP-014 standards, is to prevent attacks by monitoring critical facilities to identify would-be attackers before they attempt an attack, preventing attacker access to critical assets, and otherwise hardening facilities to make them more physically secure to protect against attack and equipment failure. The second approach is to make the broader power system more "resilient" to a successful attack on particular assets through an enhanced ability to manage loads, reroute power flows, and access other sources of generation to reduce the potential of blackouts even if critical assets are disabled.[104] Initiatives such as the spare transformer program administered by the Edison Electric Institute (EEI, the electric utility trade association), and a proposed federal Strategic Transformer Reserve, which can accelerate replacement of critical transformers if they are damaged, may contribute to the power grid's ability to sustain a terrorist attack without widespread grid failure.[105] Thus, while hardening is aimed more at reducing the likelihood of a successful attack, resilience aims at reducing potential consequence; doing either reduces overall security risk.

Measures to harden critical facilities and measures to increase system resilience are not exclusive of one another. In fact, they can be complementary in reducing overall security risk. However, they may involve different approaches to power grid operation and design, and they may involve different, competing types of investment (e.g., transformer shielding vs. transmission network sensors). Balancing the two approaches to most efficiently achieve a desired level of physical security is a challenge for utilities with limited capital budgets. The CPUC stated that "determining appropriate security measures or approaches to ensuring resiliency" was one of three "major issues" in its power grid physical security proceedings.[106] As Congress continues its oversight of bulk power physical security

[104] For a discussion about power grid resiliency and associated federal efforts, see *Government Accountability Office, Electricity: Federal Efforts to Enhance Grid Resilience*, GAO-17-153, January 2017.

[105] For details about electric sector spare transformer programs, see Department of Energy, *Strategic Transformer Reserve*, report to Congress, March 2017.

[106] CPUC, January 2018, p. 5.

regulation, it may consider whether the electric power sector as a whole is striking an appropriate balance between these two approaches.

Threat Information

The utility industry's physical security risk assessments rely upon threat information from the federal government, among other sources.[107] The quality of this threat information is a key determinant of what bulk power asset owners need to be protecting against and what security measures to take. Incomplete or ambiguous threat information may lead to inconsistency in physical security among grid owners, inefficient spending of limited security resources at facilities (e.g., that may not really be under threat), or deployment of security measures against the wrong threat.

As discussed earlier in this chapter, the E-ISAC plays a valuable role in identifying and analyzing physical security risk, and disseminating information about those risks to bulk power asset owners. Independent third-party verification of risk assessments under the CIP-014 standards, together with NERC compliance audits, are two additional means of helping to ensure greater consistency of threat information among utilities. Nonetheless, a changing threat environment continues to pose challenges for physical security planning and investment. As NERC stated in a recent compliance report, "the security threat landscape is constantly changing and requires adaptation and information sharing on how best to address these issues in an effective and efficient manner."[108]

---

[107] Much of this information is communicated primarily through the Electricity Sector Information Sharing and Analysis Center (ES-ISAC), the sector's communications channel for security-related information, situational awareness, incident management, and coordination. The ES-ISAC was established under Presidential Decision Directive 63, May 22, 1998. The ES-ISAC is operated by NERC in collaboration with the DOE and Electricity Subsector Coordinating Council. Members may anonymously share information by means of a secure Internet portal. Registered users receive information on security threats and alerts, remediation, task forces, events, and other security-specific resources.

[108] NERC, *Compliance Monitoring and Enforcement Program Quarterly Report, Q3 2017*, November 8, 2017, p. 8.

Concerns about the quality and specificity of federal threat information have long been an issue across all critical infrastructure sectors.[109] Threat information continues to be an uncertainty in the case of power grid physical security. For example, although there is wide consensus that the Metcalf attack was extremely alarming, some industry analysts have opined that FERC's physical security order nonetheless may have been an "overreaction" to Metcalf.[110] By contrast, former DHS Secretary Michael Chertoff has predicted that "the sophistication and resulting damage of the Metcalf attack will ... be exceeded" in a future attack.[111] Still others have expressed concern that FERC's physical security concerns may be too heavily focused on another Metcalf-type scenario—the last threat—rather than a wider range of potential future threats.

As discussed earlier, there is widespread belief that bulk power critical assets are vulnerable to physical attack, that such an attack potentially could have catastrophic consequences, and that the risks of such attacks are growing. But the exact nature of such potential attacks and the capability of perpetrators to successfully execute them are uncertain. Consequently, despite the technical arguments, with limited information about potential targets and attacker capabilities, the true vulnerability of the grid remains an open—and evolving—question. As Congress seeks to establish the best policies to address bulk power physical security, it may examine how federal and electric sector threat information is developed and used by critical asset owners, and how limitations and uncertainty of this information may affect physical security of the electric grid.

---

[109] See, for example, Philip Shenon, "Threats and Responses: Domestic Security," *New York Times*, June 5, 2003, p. A15.

[110] Deborah Carpentier, "NERC Gains in Vegetation Management, Cyber and Physical Security, and Reliability Assurance," *Natural Gas & Electricity* (Wiley Periodicals), May 2014, p. 31, http://www.crowell.com/files/NERCGains-in-Vegetation-Management-Cyber-and-Physical-Security-and-Reliability-Assurance.pdf.

[111] Michael Chertoff, "Building a Resilient Power Grid," *Electric Perspectives*, May/June 2014, p. 35.

CONCLUSION

The 2013 attack on the Metcalf transformer substation marked a turning point for the U.S. electric power sector. The attack prompted utilities across the country to reevaluate and restructure their physical security programs. It also set in motion proceedings in Congress and at FERC which resulted in the promulgation of NERC's CIP-014 mandatory physical security standards in 2015. Based on discussions with FERC and NERC staff about utility compliance, as well as a review of public information about the activities of bulk power asset owners (and the vendors supplying them), there appear to be physical security improvements underway among owners of bulk power critical assets. The public record is too anecdotal to assert conclusively that these changes are occurring uniformly and at every relevant utility, but NERC's summary compliance reports so far have been positive, especially for such a new standard. As NERC concluded in its *State of Reliability 2017* report,

> What NERC can measure is that no major cyber- and few physical-related load losses have happened to date; that extremely low numbers of incidents have occurred on the operating side, and that attention to security performance has been excellent on the corporate side.[112]

Although the electric power sector seems to be moving in the direction of more extensive physical security, many measures have yet to be implemented and the process of corporate realignment around physical security is still underway. As the CPUC has stated,

> It appears that the North American electric industry is in intermediate stages of fully harnessing the potential of security technologies and staff expertise, and integrating security and risk assessment values into the utility culture such that utility physical security ultimately is prioritized on par with safety and reliability.[113]

---

[112] NERC, June 2017, p. 59.
[113] CPUC, January 2018, p. 57.

Therefore, although it is probably accurate to conclude that, based on the objectives of the CIP014 standards, the U.S. electric grid is more physically secure than it was five years ago, it has not necessarily reached the level of physical security needed based on the sector's own assessments of risk. Bulk power physical security remains a work in progress. As CIP-014 implementation and other physical security initiatives proceed, Congress may seek to maintain its focus on the power sector's overall progress, not only on short term compliance with NERC's security standards, but also on structural changes supporting physical security as a priority far into the future.

In: Energy Policies: Issues and Legislation ISBN: 978-1-53614-485-7
Editor: Doug Higgins

Chapter 5

ENERGY TAX PROVISIONS THAT EXPIRED IN 2017 ("TAX EXTENDERS")*

Molly F. Sherlock, Donald J. Marples and Margot L. Crandall-Hollick

SUMMARY

Thirteen temporary energy tax provisions expired at the end of 2017. All of these provisions had expired at the end of 2016, and were retroactively extended by the Bipartisan Budget Act of 2018 (BBA; P.L. 115-123) and made available for the 2017 tax year.

This chapter briefly summarizes and discusses the economic impact of energy-related tax provisions that expired at the end of 2017, including the following:

- *Renewable energy property provisions*
  - Production Tax Credit (PTC) for Non-Wind Facilities
- *Alternative and renewable fuels provisions*
  - Incentives for Biodiesel and Renewable Diesel
  - Incentives for Alternative Fuel and Alternative Fuel Mixtures

* This is an edited, reformatted and augmented version of a Congressional Research Service report, R44990, dated March 8, 2018.

  - Alternative Fuel Vehicle Refueling Property
  - Second Generation (Cellulosic) Biofuel Producer Credit
  - Special Depreciation Allowance for Second Generation (Cellulosic) Biofuel Plant Property
- *Vehicles provisions*
  - Alternative motor vehicle credit for qualified fuel cell vehicles
  - Credit for two-wheeled plug-in electric vehicles
- *Building energy efficiency provisions*
  - Credit for Construction of Energy-Efficient New Homes
  - Energy-Efficient Commercial Building Deduction
  - Credit for Section 25C Nonbusiness Energy Property
- *Other provisions*
  - Special Rule to Implement Electric Transmission Restructuring
  - Credit for Production of Indian Coal

This chapter does not include provisions that in the past have been classified as individual or business related.

INTRODUCTION

In the past, Congress has regularly acted to extend expired or expiring temporary tax provisions.[1] Collectively, these temporary tax provisions are often referred to as "tax extenders." Of the 33 expiring provisions addressed in the most recent tax extender legislation, included in the Bipartisan Budget Act of 2018 (BBA18; P.L. 115-123), 18 were energy-related. Thirteen of the energy tax provisions addressed in BBA18 were provisions that had expired in 2016 and were extended through 2017. Thus, even though these provisions were recently extended, they remain expired. These 13 expired energy tax provisions are reviewed in this chapter.

The expired energy tax provisions are diverse in purpose, providing various types of tax incentives for renewable electricity, alternative and renewable fuels, vehicles, building energy efficiency, and other energy activities. Before BBA18, expired energy tax provisions were extended in the Protecting Americans from Tax Hikes (PATH) Act of 2015, enacted as

---

[1] For an overview of tax extenders, see CRS Report R44677, *Tax Provisions that Expired in 2016 ("Tax Extenders")*, by Molly F. Sherlock.

Division Q of the Consolidated Appropriations Act, 2016 (P.L. 114-113). This law retroactively extended temporary tax provisions that had expired at the end of 2014. Unlike much of the previous tax extender legislation, this law did not temporarily extend most or all expired provisions. Instead, a number of temporary provisions were made permanent, while others were extended through 2016 or 2019. While certain business, individual, and charitable provisions were made permanent or extended through 2019, energy-related provisions were generally extended for two years, through 2016, in the PATH Act.

There were longer-term extensions for certain renewables in Division P of P.L. 114-113. Specifically, the law provided longer-term extensions with scheduled phaseouts of tax credits for wind and solar. The renewable electricity production tax credit (PTC) for wind was extended through 2019, with the phaseout starting for facilities beginning construction in 2017.[2] For wind facilities that begin construction during 2017, the credit is reduced by 20%. The credit is reduced by 40% for facilities that begin construction in 2018, and reduced by 60% for facilities that begin construction in 2019.

The 30% investment tax credit (ITC) for business solar was extended through 2019 and the deadline changed from a placed-in-service deadline to a construction start date.[3] The business solar ITC was set to be 26% for facilities beginning construction in 2020 and 22% for facilities beginning construction in 2021, so long as these facilities are placed in service before the end of 2023. The business solar ITC is scheduled to return to 10% in 2022. The tax credit for residential solar was extended through 2021, with a phaseout starting in 2020.

Several provisions in BBA18 provided extensions for energy-tax provisions that matched the longer-term extensions that had been enacted as part of the PATH Act. Specifically, the ITC for non-solar technologies was extended and modified, such that the phase-out and termination for non-solar

---

[2] CRS Report R43453, *The Renewable Electricity Production Tax Credit: In Brief*, by Molly F. Sherlock.

[3] CRS In Focus IF10479, *The Energy Credit: An Investment Tax Credit for Renewable Energy*, by Molly F. Sherlock.

technologies matched what had been established for solar in the PATH Act.[4] Five-year cost recovery that is generally available to ITC-eligible property, as well as wind property, was also extended through 2021. The tax credit for residential energy-efficient property was also extended through 2021 for non-solar technologies, with a phase-out in 2020 and 2021.[5] Further discussion of these provisions is not included in this chapter, as these provisions did not expire in 2017.

Additional energy tax provisions were included in BBA18. Specifically, the law extended the per-barrel excise tax on oil used to fund the oil spill liability trust fund,[6] modified the tax credit for production from advanced nuclear power facilities,[7] and enhanced the carbon dioxide sequestration credit.[8] Again, further discussion of these provisions is not included in this chapter, as these provisions did not expire in 2017.

There are several options for Congress to consider regarding temporary provisions. The 2017 tax revision (P.L. 115-97) adds new context in considering further extension of expired temporary tax provisions.[9] Broadly, the revised tax system became effective in 2018. Thus, one perspective might be that extending expired provisions through 2017 allows these provisions to remain available up until the revised tax system took effect at the beginning of 2018. Whether the temporary tax provisions that expired at the end of 2017 have a place in the new tax system is a question the 115th Congress may consider. Ultimately, provisions that

---

[4] The credit is extended through 2021 for all qualifying technologies. For fiber optic solar energy property, qualified fuel cell property, and qualified small wind property, the credit rate is 30% for property that begins construction before the end of 2019, falling to 26% in 2020 and 22% in 2021. Additionally, property must also be placed in service by the end of 2023 to qualify for the tax. The 10% credit for qualified microturbine property, combined heat and power (CHP) property, and thermal energy property is also extended through 2021.

[5] The credit was extended for solar water heating property, fuel cell plants, small wind energy property, and geothermal heat pump property. The credit rate is 30% for property placed in service before the end of 2019, falling to 26% in 2020 and 22% in 2021.

[6] CRS In Focus IF10823, *The Oil Spill Liability Trust Fund Tax: Reauthorization Issues and Legislation in the 115th Congress*, by Jonathan L. Ramseur.

[7] CRS Insight IN10725, *The Advanced Nuclear Production Tax Credit*, by Molly F. Sherlock and Mark Holt.

[8] The Section 45Q tax credit is expanded to apply to all carbon oxides for facilities placed into service on or after February 9, 2018, which begin construction by the end of 2023. Credit amounts are also increased for facilities placed in service after February 9, 2018.

[9] CRS Report R45092, *The 2017 Tax Revision (P.L. 115-97): Comparison to 2017 Tax Law*, coordinated by Molly F. Sherlock and Donald J. Marples.

expired at the end of 2017 could be extended. The extension could be retroactive, such that provisions will be available for all of the 2018 tax year. The extensions also could be short term, long term, or permanent. Another option would be to allow expired provisions to remain expired.

Table 1 provides information on (1) the cost of the one-year extension of expired energy tax provisions in BBA18, over the 2018 to 2027 budget window; and (2) the estimated reduction in federal revenues that would result from making expired energy tax provisions permanent.

RENEWABLE ELECTRICITY

Production Tax Credit (PTC) for Non-Wind Facilities[10]

The renewable electricity PTC is a per-kilowatt-hour (kWh) credit for electricity produced by a qualified energy resource. Eligible facilities can claim the tax credit for the first 10 years of qualified production. Resources that qualify for the full credit amount, 2.4 cents per kWh in 2017, include wind, closed-loop biomass, and geothermal.

Other resources are eligible for a half credit amount, including open-loop biomass, small irrigation power, municipal solid waste, qualified hydropower, and marine and hydrokinetic resources. Under current law, non-wind facilities must have begun construction before January 1, 2017, to be tax-credit eligible.[12]

The PTC was enacted in 1992 as part of the Energy Policy Act (EPACT92; P.L. 102-486). When first enacted, the PTC was available for electricity generated using wind or closed-loop biomass systems. The credit was initially set to expire on June 30, 1999. Since 1999, the PTC has regularly been extended, often as part of "tax extenders" legislation.

---

[10] Internal Revenue Code (IRC) Section 45(d). IRC Section 48(a)(5) for the investment tax credit (ITC) in lieu of PTC option.

[12] Wind facilities beginning construction after December 31, 2016, may qualify for the PTC at reduced rates. Wind facilities that begin construction after December 31, 2019, will not be eligible for the PTC.

Table 1. Energy Tax Provisions That Expired at the End of 2017

	10-Year Cost Estimate of 1-Year Extension in P.L. 115-123 (billions)	Cost of Permanent Extension—CBO Policy Alternatives Baseline (Before the P.L. 115-97) (billions)
Renewables		
Production Tax Credit (PTC) for Non-Wind Facilities	$0.3	$2.8
Alternative and Renewable Fuels		
Incentives for Biodiesel and Renewable Diesel	$3.3	$37.5
Incentives for Alternative Fuel and Alternative Fuel Mixtures	$0.6	$7.4
Alternative Fuel Vehicle Refueling Property	$0.1	$1.4
Second Generation (Cellulosic) Biofuel Producer Credit	-i-	$0.3
Special Depreciation Allowance for Second Generation	-i-	$0.0
(Cellulosic) Biofuel Plant Property		
Vehicles		
Alternative Motor Vehicle Credit for Qualified Fuel Cell	-i-	$0.1
Vehicles		
Credit for Two-Wheeled Plug-in Electric Vehicles	-i-	$0.0
Building Energy Efficiency		
Credit for Construction of Energy-Efficient New Homes	$0.3	$3.5
Energy-Efficient Commercial Building Deduction	$0.1	$2.1
Credit for Section 25C Nonbusiness Energy Property	$0.5	$6.5
Other		
Special Rule to Implement Electric Transmission Restructuring	—	$0.0
Credit for Production of Indian Coal	-i-	$0.4

Source: CRS analysis of extending legislation; Joint Committee on Taxation, List of Expiring Federal Tax Provisions 2016-2027, January 9, 2018, JCX-1-18; Congressional Budget Office (CBO), An Update to the Budget and Economic Outlook, Detailed Revenue Projections, June 2017; and Joint Committee on Taxation, Estimated Budget Effects of the Revenue Provisions Contained in the "Bipartisan Budget Act of 2018," 115th Cong., February 8, 2018, JCX-4-18.

Notes: The cost of permanent extension is as reported by CBO for the 2018 to 2027 budget window. These estimates were prepared before the 2017 tax revision (P.L. 115-97) was enacted. Provisions enacted as part of tax reform, such as the reduced corporate tax rate, may cause changes in the estimated cost of extending certain provisions. An -i- indicates an estimated revenue loss of less than $50 million between 2018 and 2027. An "n/a" means consideration of extension was not applicable, for reasons explained in the text. A "—" means JCT estimated no revenue effect.

The credit has also been expanded to include additional qualifying resources. At several points in time, the PTC was allowed to lapse before a retroactive extension was enacted.

Taxpayers can elect to receive a 30% investment tax credit in lieu of the PTC. The ITC in lieu of PTC election was enacted in 2009 alongside the Section 1603 grants in lieu of tax credits that expired at the end of 2011.[13] In recent years, the option to elect the ITC in lieu of the PTC has been extended alongside the PTC in "tax extenders" legislation.

The PTC was enacted in 1992 to promote the "development and utilization of certain renewable energy sources."[14] The 1999 sunset was included to provide an "opportunity to assess the effectiveness of the credit."[15] When the PTC was extended as part of a "tax extenders" package in 1999, Congress noted that the PTC had been important to the development of environmentally friendly renewable power, and extended the credit to promote further development of wind (and other) resources.[16] Subsequent extensions of the PTC reflected a belief that the tax incentive contributed to the development of renewable-energy infrastructure, which advanced environmental and energy policy goals.[17]

Most recently, the PTC was extended for one year for qualifying non-wind technologies, from 2016 through 2017, in BBA18. Previously, a two-year extension of the PTC for non-wind technologies was included in the Protecting Americans from Tax Hikes (PATH) Act of 2015, enacted as Division Q of P.L. 114-113. Division P of the Consolidated Appropriations Act, 2016 (P.L. 114-113) included a two-year extension of the PTC for wind (through 2016). The PTC for wind was also extended beyond 2016, through 2019, at reduced rates.

---

[13] CRS Report R41635, ARRA Section 1603 Grants in Lieu of Tax Credits for Renewable Energy: Overview, Analysis, and Policy Options, by Phillip Brown and Molly F. Sherlock.

[14] U.S. Congress, House Committee on Ways and Means, Comprehensive National Energy Policy Act, committee print, 102nd Cong., 2nd sess., May 5, 1992, H. Rept. 102-474, pp. 41-42.

[15] Ibid.

[16] U.S. Congress, Joint Committee on Taxation, *General Explanation of Tax Legislation Enacted in the 106th Congress*, committee print, April 19, 2001, JCS-2-01, p. 25.

[17] U.S. Congress, Joint Committee on Taxation, *General Explanation of Tax Legislation Enacted in the 112th Congress*, committee print, February 2013, JCS-2-13, pp. 212-213.

One policy rationale for supporting renewable electricity resources is to promote development of renewable-energy infrastructure that may not be technologically mature. With this policy rationale in mind, one question is when technologies have reached the point of maturity such that tax-related federal financial support can be eliminated.

Environmental considerations provide another policy rationale often used to support tax incentives for renewable electricity production. Some suggest that ongoing support for renewables can help address inefficiencies and market failures in energy production markets, where fossil-fuels-based electricity production and the associated pollution effects generate negative externalities. A more direct approach, however, would be to impose a price on pollution, as opposed to subsidizing a nonpolluting alternative.

ALTERNATIVE AND RENEWABLE FUELS

Incentives for Biodiesel and Renewable Diesel[18]

There are three tax credits for biodiesel: the biodiesel mixture credit, the biodiesel credit, and the small agri-biodiesel producer credit. Each gallon of biodiesel, including agri-biodiesel (biodiesel made from virgin oils), may be eligible for a $1.00 tax credit. The mixtures tax credit may be claimed as an instant excise tax credit against the blender's motor and aviation fuels excise taxes. Credits in excess of excise tax liability may be refunded. The biodiesel and small agri-biodiesel credits may be claimed as income tax credits. The mixtures credit is proportionate to the fraction of biodiesel in the mixture—a blend of 80% diesel with 20% virgin biodiesel would qualify for a 20-cent-per-gallon tax credit. The tax credits for biodiesel expired on December 31, 2017.

Additionally, an eligible small agri-biodiesel producer credit of 10 cents is available for each gallon of "qualified agri-biodiesel production." An eligible "small agri-biodiesel producer" is defined as any person who,

[18] IRC Sections 40A, 6426(c)(6), and 6427(e)(6)(B).

at all times during the taxable year, has annual productive capacity for agri-biodiesel not in excess of 60 million gallons. The number of gallons that may be taken into account for the small agri-biodiesel producer credit may not exceed 15 million. The eligible small agri-biodiesel producer credit is effective for taxable years ending after August 8, 2005, and sunsets after December 31, 2017.

The tax code generally treats renewable diesel fuel like biodiesel for the purposes of the biodiesel fuels credit. Thus, renewable diesel sold or used after December 31, 2005, is eligible for a $1.00 per gallon tax credit. Renewable diesel cannot qualify as agri-biodiesel.

The biodiesel tax incentives were introduced under the American Jobs Creation Act of 2004 (P.L. 108-357) and modified and extended by the Energy Policy Act of 2005 (P.L. 109-58). Subsequently, the credits have been extended, most recently through the end of 2017 in BBA2018.

Tax credits for biofuels are motivated by a desire to reduce dependence on petroleum imports (enhance national energy security), address environmental concerns, and maintain farm incomes.

While the use of biofuels continues to increase, offsetting domestic petroleum consumption, it is not clear that the tax incentives are responsible for driving this change. Renewable fuel standards and blend mandates requiring certain amounts of biofuels may be boosting domestic production.[19] If nontax policies are responsible for enhancing biofuel production, and tax policies fail to induce additional production, the tax credits provide a windfall to taxpayers without necessarily resulting in additional use of biofuels.

Incentives for Alternative Fuel and Alternative Fuel Mixtures[20]

The tax code provides tax credits for alternative fuels and alternative fuel mixtures. Specifically, there is a 50-cents-per-gallon excise tax credit for certain alternative fuels used as fuel in a motor vehicle, motor boat, or

---

[19] CRS Report R43325, *The Renewable Fuel Standard (RFS): An Overview*, by Kelsi Bracmort.
[20] IRC Sections 6426(d)(5), 6427(e)(6)(C), and 6426(e)(3).

airplane and a 50-cents-per-gallon credit for alternative fuels mixed with a traditional fuel (gasoline, diesel, or kerosene) for use as a fuel. Qualifying fuels include liquefied petroleum gas; P Series fuels (certain renewable, nonpetroleum, liquid fuels); compressed or liquefied natural gas (CNG or LNG); any liquefied fuel derived from coal or peat through the Fischer-Tropsch process which meets certain carbon-capture requirements; liquefied hydrocarbons derived from biomass; and liquefied hydrogen. For propane, CNG, and LNG sold after December 31, 2015, the tax credit is based on gasoline-gallon or diesel-gallon equivalent. No fuel produced outside of the United States is eligible for the alternative fuels tax incentives.

The incentives of alternative fuel and alternative fuel mixtures were introduced under the Safe, Accountable, Flexible, Efficient Transportation Equity Act: A Legacy for Users (SAFETEA-LU; P.L. 109-59). Initially the credits were to be available through September 30, 2009 (September 30, 2014, for hydrogen). The credit, however, has subsequently been extended. Most recently the incentives were extended through the end of 2017 in BBA2018.

Tax credits for alternative fuels are motivated by a desire to reduce dependence on petroleum imports (enhance national energy security), address environmental concerns, and maintain farm incomes.

Alternative Fuel Vehicle Refueling Property[21]

A 30% tax credit is provided for the cost of any qualified alternative fuel vehicle refueling property installed by a business or at a taxpayer's principal residence. The credit is limited to $30,000 for businesses at each separate location, and $1,000 for residences.

Clean fuel refueling property is generally any tangible equipment (such as a pump) used to dispense a fuel into a vehicle's tank. Qualifying property includes fuel storage and dispensing units and electric vehicle

[21] IRC Section 30C(g).

recharging equipment. A clean fuel is defined as any fuel at least 85% of the volume of which consists of ethanol (E85) or methanol (M85), natural gas, CNG, LNG, liquefied petroleum gas, and hydrogen, or any mixture of biodiesel and diesel fuel, determined without regard to any use of kerosene and containing at least 20% biodiesel. For the purposes of the credit, electricity is also considered a clean fuel.

For business taxpayers, the taxpayer's basis in the property is reduced by the amount of the credit. The credit for business property is treated as a portion of the general business credit. As part of the general business credit, unused credits may be carried back for one year or carried forward for 20 years. For nonbusiness property, the credit cannot exceed the excess of an individual's income tax liability over the sum of nonrefundable personal credits and the foreign tax credit over the taxpayer's tentative minimum tax. No credit is available for property used outside the United States. For property sold to a tax-exempt entity, the seller of the property may be able to claim the credit.

The credit for alternative fuel vehicle refueling property was introduced under the Energy Policy Act of 2005 (EPACT05; P.L. 109-58). The credit replaced a previously available deduction for business investment in clean fuel refueling property. The credit has been extended multiple times since being enacted in 2005; most recently, the incentives were extended through the end of 2017 in BBA2018.

Tax credits for alternative fuel vehicle refueling property reduce after-tax capital costs to attract investment. Additionally, nontax federal incentives may also promote investment in alternative fuel vehicle refueling property. From an economic perspective, allowing special tax credits for selected technologies can distort the allocation of resources, and may create economic inefficiencies by encouraging investments in high-cost technologies, ones that would not otherwise be economical at current and expected prices and rates of return. However, the incentive may improve the allocation of resources, if it corrects a market failure.

Second Generation (Cellulosic) Biofuel Producer Credit[22]

The second generation biofuels producer credit is a nonrefundable income tax credit for each gallon of qualified second generation biofuel production. The amount of the credit per gallon is generally $1.01. Qualified second generation biofuel production is second generation biofuel produced by the taxpayer and sold by the taxpayer to another person for use (1) in the production of a qualified biofuel fuel mixture in such person's trade or business (other than casual off-farm production), (2) as a fuel in a trade or business, or (3) as biofuel sold at retail to another person and placed in the fuel tank of such other person. Qualifying fuel must be produced and used in the United States. Cellulosic biofuel is produced using lignocellulosic or hemicellulosic matter available on a renewable or recurring basis. Qualified feedstocks for second generation biofuels include cultivated algae, cyanobacteria, or lemna.

The cellulosic biofuel producer credit was introduced by the Food, Conservation, and Energy Act of 2008 (P.L. 110-246). When introduced, the credit was scheduled to expire December 31, 2012. The American Taxpayer Relief Act of 2012 (ATRA; P.L. 112-240) modified the provision to algae-based fuels, changing the title of the provision to be the second generation biofuel producer credit at the same time. It has subsequently been extended as part of "tax extenders." Most recently the incentive was extended through the end of 2017 in BBA2018.

Tax credits for second generation biofuels are motivated by a desire to reduce dependence on petroleum imports (enhance national energy security), address environmental concerns, and maintain farm incomes. Renewable fuel standards and blend mandates requiring certain amounts of biofuels may be boosting domestic production, but until recently second generation cellulosic biofuels were not economically competitive with other renewable fuel standards options.[23]

---

[22] IRC Section 40(b)(6)(J).

[23] CRS Report R43325, *The Renewable Fuel Standard (RFS): An Overview*, by Kelsi Bracmort.

Special Depreciation Allowance for Second Generation (Cellulosic) Biofuel Plant Property[24]

Second generation biofuel plant property allows for the immediate first-year bonus depreciation of 50% of the cost of facilities that produce eligible biofuels. Previous federal tax law limited the eligibility for first-year bonus depreciation of cellulosic biofuels to facilities producing ethanol; those producing nonethanol fuels from cellulosic feedstocks did not qualify for the allowance.

The special depreciation allowance for second generation biofuel plant property was introduced by the Tax Relief and Health Care Act of 2006 (P.L. 109-432). Most recently the incentive was extended through the end of 2017 in BBA2018. Allowing half the cost to be expensed when incurred provides a benefit because a tax deduction today is worth more than a tax deduction in the future, due to the time value of money (interest). From an economic perspective, allowing a special depreciation allowance for selected technologies can distort the allocation of resources, and may create economic inefficiencies. However, this incentive may also increase economic efficiency to the extent it addresses a market failure.

VEHICLES

Alternative Motor Vehicle Credit for Qualified Fuel Cell Vehicles[25]

Fuel cell vehicles (vehicles propelled by chemically combining oxygen with hydrogen to create electricity) may qualify for a federal tax credit. The credit is based on vehicle weight, with vehicles weighing 8,500 pounds or less having a base credit amount of $4,000. Heavier vehicles may be eligible for larger credits, with the highest credit amount being

---

[24] IRC Section 168(l).
[25] IRC Section 30B(k)(1).

$40,000 for vehicles weighing more than 26,000 pounds. Cars and light trucks can qualify for an additional tax credit of $1,000 to $2,000 per vehicle depending on fuel economy.

The alternative technology vehicle tax credit was enacted as part of the Energy Policy Act of 2005 (P.L. 109-58). The credit replaced a previously available clean-fuel-vehicle deduction (IRC Section 179A). When enacted, the tax credit for fuel cell vehicles was available for vehicles placed in service during 2006 through 2014. Alternative technology vehicle tax credits were available for hybrid, advanced lean-burn technology, and alternative fuel vehicles. The credits generally expired in 2009 or 2010 for vehicles other than fuel cell vehicles. After 2014, the credit for fuel cell vehicles has been extended as part of "tax extenders." Most recently, the provision was extended through 2017 in BBA2018.

Tax incentives for fuel cell vehicles may help address market failures in automobile markets. Specifically, if consumers fail to consider the negative environmental and potential energy security concerns associated with conventional gasoline- and diesel-fueled vehicles, the market may provide an inefficiently high level of such products. One way to address the negative externalities associated with fuel consumption through automobile use is to reduce the price of alternative technology vehicles.

There are other barriers to adoption of alternative-technology vehicles a tax credit might address. These include, for example, (1) the high up-front cost associated with alternative-technology vehicles, (2) the volatility of fuel prices, (3) technology risks associated with new, unfamiliar, or unproven technologies, and (4) a lack of complementary infrastructure (such as hydrogen fueling stations).

Because tax credits for fuel cell vehicles reduce the price of such vehicles relative to gasoline- and diesel-powered alternatives, such tax credits are intended to address the previously noted market failures and market barriers. A tax credit approach, however, may not be the most economically efficient mechanism for addressing the negative externalities associated with gasoline consumption and market barriers to fuel cell vehicle adoption. Relative to tax credits, rising gas prices have played a larger role in increasing consumer demand for alternative technology vehicles. Taxing gasoline directly—the activity associated with the negative externality—is arguably more economically efficient than subsidizing the purchase of select vehicles.

Credit for Two-Wheeled Plug-In Electric Vehicles[26]

A credit is available for two-wheeled plug-in vehicles acquired during 2015 or 2016. The credit is equal to 10% of the vehicle's cost, up to $2,500. To be eligible for the tax credit vehicles must have a weight rating of less than 14,000 pounds; be propelled by a battery-powered electric motor with a battery capacity of at least 2.5 kilowatt hours; be manufactured for use on streets, roads, and highways; and be capable of achieving a speed of at least 45 miles per hour.

The American Recovery and Reinvestment Act (ARRA; P.L. 111-5) provided a tax credit for two- or three-wheeled vehicles, as well as low-speed four-wheeled vehicles. When enacted, the provision was scheduled to expire at the end of 2011. In ATRA, the provision was extended through 2013 for two- and three-wheeled vehicles, but not low-speed vehicles. No credit was available in 2014. The provision was reinstated for two-wheeled vehicles in the PATH Act of 2015 (Division Q of P.L. 114-113), and extended through 2017 in BBA2018.

Credits for two-wheeled plug-in electric vehicles were enacted as a unique provision, since they do not qualify for other plug-in electric vehicle tax credits.[27] Tax credits for such vehicles can support emerging technologies, or encourage consumers to purchase vehicles that might be more energy efficient than conventional alternatives. Some have raised concerns regarding the equity of vehicles tax benefits, noting that they tend to be largely claimed by high-income taxpayers. If these taxpayers would have bought qualifying vehicles absent tax benefits, then the tax incentives are not leading to additional purchases and are providing a windfall benefit to purchasers.

---

[26] IRC Section 30D(g)(3)(E)(ii).

[27] See the new qualified plug-in electric-drive vehicle credit, IRC Section 30D.

BUILDING ENERGY EFFICIENCY

Credit for Construction of Energy-Efficient New Homes[28]

Before 2017, contractors building energy-efficient homes and producers of manufactured energy-efficient homes were eligible for a tax credit for each qualifying new home they built. The amount of the credit was equal to $2,000 per home for homes built by contractors and $1,000 per manufactured home.

To be eligible, an energy-efficient new home was required to have annual heating and cooling consumption that was at least 50% (30% in the case of manufactured homes) below a comparable unit. The home was also required to be in accordance with the standards of the 2006 International Energy Conservations Code.[29] Contractors and manufacturers claiming this credit were required to submit certification to an eligible certifier before claiming the credit. This credit was part of the general business credit and hence could be carried back one year and carried forward 20 years.

The tax credit for energy-efficient new homes was introduced under the Energy Policy Act of 2005 (P.L. 109-58). Initially, the credit was set to expire at the end of 2007. It was subsequently extended several times.[30] Most recently, the credit was extended through the end of 2017 in BBA2018. The tax credit is designed to encourage home builders to install energy-efficient technologies in new homes. Energy-efficient new homes

---

[28] IRC Section 45L(g).

[29] In addition, heating and cooling equipment efficiency must correspond to the minimum allowed under the regulations established by the Department of Energy pursuant to the National Appliance Energy Conservation Act of 1987 (P.L. 100-12) in effect at the time construction is completed. Qualified homes must be constructed such that building-envelope components contribute at least a fifth of the 50% in required energy consumption reduction (1/3 of 30% in required energy consumption reduction in the case of manufactured homes). Energy Star labeled homes may qualify for the tax credit.

[30] The Tax Relief and Health Care Act of 2006 (P.L. 109-432) extended the credit through December 31, 2008. The Emergency Economic Stabilization Act of 2009 (EESA; P.L. 110-343) extended the credit through December 31, 2009. The Tax Relief, Unemployment Insurance Reauthorization and Job Creation Act of 2010 (P.L. 111-312) extended the credit through December 31, 2011. The American Taxpayer Relief Act of 2012 (P.L. 112-240) extended the credit until December 31, 2013, and adopted the 2006 International Energy Conservation Code.

tend to have higher up-front costs, and it is not clear if market prices accurately reflect or capitalize the value of energy-efficient improvements. If energy efficiency is not accurately reflected in housing prices, builders may underinvest in efficiency technologies absent the credit. On the other hand, if market prices do reflect the cost associated with these technologies, the credit may be a windfall gain to the builder.

Energy-Efficient Commercial Building Deduction[31]

A deduction is allowed for certain energy-saving property used in domestic commercial buildings.[32] Qualifying energy-efficient commercial building property includes property installed as part of (1) the interior lighting system; (2) the heating, cooling, ventilation, or hot water system; or (3) the building envelope. To be deductible, property must reduce a building's annual energy and power costs by 50% or more as compared to a similar reference building meeting the minimum energy standards described in Standard 90.1-2007 of ASHRAE/IESNA.[33] The maximum deduction allowed is $1.80 per square foot. A reduced deduction may be available if a single system is upgraded (lighting, heating and cooling, or building envelope) and the 50% reduction threshold is not met. Separate energy cost reduction percentage thresholds are specified for single-system upgrades. The maximum deduction for a single-system improvement is $0.60 per square foot. Government entities making energy-efficiency upgrades to public buildings, such as schools, can allocate the Section 179D deduction to designers of energy-efficient commercial building property. The deduction for energy-efficient commercial building property was enacted in the Energy Policy Act of 2005 (EPACT05; P.L. 109-58). When first enacted, the deduction was scheduled to be available for the

---

[31] IRC Section 179D(h).

[32] For a more detailed overview, see CRS Committee Print CP10002, *Tax Expenditures: Compendium of Background Material on Individual Provisions — A Committee Print Prepared for the Senate Committee on the Budget, 2016*, by Jane G. Gravelle et al., pp. 107-113.

[33] American Society of Heating, Refrigerating, and Air Conditioning Engineers and the Illuminating Engineering Society of North America standards.

2006 and 2007 tax years. The deduction was extended for one year, through 2008, in "tax extenders" legislation enacted late in 2006 (the Tax Relief and Health Care Act of 2006 [P.L. 109-432]). A longer-term (five-year) extension was enacted in the Emergency Economic Stabilization Act (P.L. 110-343). A stated rationale claimed

> [t]he Congress recognizes that a substantial portion of U.S. energy consumption is attributable to commercial buildings, and that the design and construction of commercial buildings is a multi-year process. Hence, the Congress believes that a long-term extension of the deduction for energy efficient commercial buildings is necessary to ensure that buildings currently in the design phase will be able to claim the deduction.[34]

Since 2014, short-term extensions of the deduction for energy-efficient commercial building property have been included in "tax extenders" legislation. Most recently, the provision was extended through 2017 in BBA2018. The business profit maximizing (average cost minimizing) objective should, in theory, promote an economically efficient level of investment in energy-saving property. However, market outcomes may lead to less investment in building energy efficiency than is socially desirable if (1) consumption of energy has negative external costs, such as pollution, that are not considered when building owners make energy property decisions (there are "negative externalities" associated with building energy consumption); or (2) if the person choosing the energy equipment is not the same person responsible for paying the energy bills (there is a "principal-agent" issue), and energy-savings investments cannot be recouped via higher rents or appreciated asset value. In these circumstances, federal financial assistance, through a tax incentive, for example, may improve upon market outcomes. The tax deduction may not be the most economically efficient way to address market inefficiency. If building energy consumption is associated with pollution-related negative

---

[34] U.S. Congress, Joint Committee on Taxation, *General Explanation of Tax Legislation Enacted in the 110th Congress*, committee print, March 2009, JCS-1-09, p. 344.

externalities, a direct price on pollution would arguably be a more economically efficient policy.

There are also some considerations related to the specifics of the deduction for energy-efficient commercial property. Most of the deduction is claimed by taxpayers constructing new buildings, since the energy-efficiency threshold is tied to recent efficiency standards, making it more difficult for older buildings to complete retrofits that meet energy-savings targets tied to recent building standards. Meeting certification requirements can also be costly and burdensome, potentially preventing certain taxpayers from claiming the deduction. Conversely, stringent certification requirements can help prevent fraudulent deduction claims.

Credit for Section 25C Nonbusiness Energy Property[35]

The nonbusiness energy property tax credit provides homeowners with a nonrefundable[36] tax credit for investments in both high-efficiency energy property (e.g., heating, cooling, and water-heating appliances) and investments in certain energy-efficiency improvements (e.g., energy-efficient insulation, windows, and doors).

For installations made during 2011 through 2017, the amount of the credit was calculated as 10% of expenditures on building-envelope improvements plus the cost of each energy-efficient property capped at a specific amount (ranging from $50 to $300), excluding labor and installation costs.[37] Given the price of high-efficiency heating, cooling, and water-heating appliances, taxpayers generally claimed the maximum amount of the credit for energy-efficient property. In addition, the credit was subject to a lifetime cap of $500 per taxpayer.

Residential energy-efficiency tax credits were first introduced in the late 1970s, but were allowed to expire in 1985. EPACT05 (P.L. 109-58)

---

[35] IRC Section 25C(g).

[36] Nonrefundable tax credits cannot exceed a taxpayer's income tax liability, meaning those taxpayers with little to no tax liability generally cannot claim these tax benefits.

[37] For more information on energy related caps, see CRS Report R42089, *Residential Energy Tax Credits: Overview and Analysis*, by Margot L. Crandall-Hollick and Molly F. Sherlock.

enacted the Section 25C credit as a temporary provision in effect for 2006 and 2007. This nonrefundable tax credit was equal to 10% of qualified expenditures, subject to certain limitations for specific types of property and a $500 lifetime limitation per taxpayer. At the end of 2007, the Section 25C credit expired. In 2008, the Emergency Economic Stabilization Act of 2008 (EESA; P.L. 110-343) reinstated and modified the Section 25C credit for the 2009 tax year. The American Recovery and Reinvestment Act of 2009 (ARRA; P.L. 111-5) further extended the credit for two years (2009 and 2010) as well as expanded it. Under ARRA, the credit equaled 30% of qualified expenditures for energy-efficiency improvements and energy property, eliminating the technology-specific credit amounts.[38] In addition, the lifetime credit cap was lifted from $500 to $1,500 for 2009 and 2010. These changes expired at the end of 2010 and the credit as structured *before ARRA* (10% of expenditures subject to a $500 lifetime cap) was subsequently extended several times on a temporary basis. The 25C credit was extended for two years—2015 and 2016—by the PATH Act of 2015 (Division Q of P.L. 114-113) and extended through 2017 in BBA2018.

The amount of the investment resulting from the 25C credit is unclear. Some researchers found that tax incentives that reduced the price of energy-efficiency property would lead to additional investment.[39] Others found that the tax credits were instead more likely associated with windfall gains to credit recipients as opposed to additional energy-efficiency investment.[40] Further, the fact that the incentive is delivered as a nonrefundable credit limits the provision's ability to motivate investment for low- and middle-income taxpayers with limited tax liability. The administration of residential energy-efficiency tax credits has also had compliance issues, as identified in

---

[38] The changes that ARRA made to the Section 25C credit in 2009 superseded the 2009 changes that had been made to the credit by EESA.

[39] See Kevin A. Hassett and Gilbert E. Metcalf, "Energy Tax Credits and Residential Conservation Investment: Evidence from Panel Data," *Journal of Public Economics*, vol. 57, no. 2 (June 1995), pp. 201-217.

[40] See Michael J. Walsh, "Energy Tax Credits and Housing Improvement," *Energy Economics*, vol. 11, no. 4 (October 1989), pp. 275-284 and Jeffery A. Dubin; and Steven E. Henson, "The Distributional Effects of the Federal Energy Tax Act," *Resources and Energy*, vol. 10, no. 3 (1988), pp. 191-212.

a Treasury Department Inspector General for Tax Administration (TIGTA) report.[41]

OTHER EXPIRED ENERGY TAX PROVISIONS

Special Rule to Implement Electric Transmission Restructuring[42]

IRC Section 451(i) permits taxpayers to elect to recognize any capital gain from the sale of qualifying electricity transmission property to an independent transmission company (ITC), pursuant to a Federal Energy Regulatory Commission (FERC) restructuring policy, evenly over eight years beginning with the year of the sale. The sale proceeds must be reinvested in other electricity assets within four years. This special tax incentive is available for sales made through December 31, 2017.

Generally, any gain realized from a sale or disposition of a capital asset is recognized in the tax year in which the gain was realized, unless there is a specific exemption or deferral. The recognition of gain over eight years, rather than in the year of sale, is a deferral, rather than a complete forgiveness, of tax liability. The economic benefit derives from the reduction in the present value of the tax owed below what the tax would otherwise be if it were required to be recognized in the year of sale.

The deferral of gain on the sale of transmission assets was enacted on a temporary basis as part of the American Jobs Creation Act of 2004 (P.L. 108-357), with the goal of encouraging energy transmission infrastructure reinvestment and assisting those in the industry who are restructuring. It is intended to foster a more competitive industry by facilitating the unbundling of transmission assets held by vertically integrated utilities. Under restructuring, states and Congress have considered rules requiring the separate ownership of generation and distribution and transmission

---

[41] Treasury Inspector General For Tax Administration, *Processes Were Not Established to Verify Eligibility for Residential Energy Credits*, Reference Number: 2011-41-038, April 19, 2011, http://www.treasury.gov/tigta/auditreports/2011reports/201141038fr.pdf.

[42] IRC Section 451(i).

assets. However, vertically integrated electric utilities still own a large segment of the nation's transmission infrastructure. The tax provision encourages the sale of transmission assets by vertically integrated electric utilities—the unbundling of electricity assets—to independent system operators or regional transmission organizations, who would own and operate the transmission lines. The provision is intended to improve transmission management and service, and facilitate the formation of competitive electricity markets. In recent years, this provision has been extended as part of "tax extenders" legislation. Most recently, the provision was extended through 2017 in BBA2018.

The restructuring of the electric power industry has resulted in significant reorganization of power assets, and may continue to do so. In particular, it may result in a significant disposition of transmission assets and possibly, depending on the nature of the transaction, trigger an income tax liability and interfere with industry restructuring. Under an income tax system, the sale for cash of business assets subject to depreciation deductions triggers a tax on taxable income in the year of sale to the extent of any gain. Corporations pay capital gains on sales of capital assets, such as shares of other corporations. But gains on the sale of depreciable assets involve other rules. For example, sales of personal property, such as machinery, are taxed partly as capital gains and partly as ordinary income. The overall taxable amount is the difference between the sales price and basis, which is generally the original cost minus accumulated depreciation. That amount is taxed as ordinary income to the extent of previous depreciation allowances (depreciation is "recaptured").

Credit for Production of Indian Coal[43]

The credit for Indian coal production provides a tax credit for Indian coal produced from reserves, which on June 14, 2005, were owned by an Indian tribe or held in trust by the United States for a tribe. The amount of the credit is $2.00 per ton (adjusted for inflation; $2.387 per ton in 2016).

---

[43] IRC Section 45(e)(10)(A).

The credit is available for coal sold after December 31, 2005, and before January 1, 2018. The coal does not need to be sold for the production of electricity or any specific purpose. Before 2015, the credit was only available for Indian coal produced at facilities that were placed in service before January 1, 2009.

Indian coal was added to the production tax credit (PTC) as a new qualifying resource by the Energy Policy Act of 2005 (P.L. 109-58). When enacted, the credit was available for coal sold to an unrelated third party during the seven-year period beginning after December 31, 2005, and ending before January 1, 2013. The credit was $1.50 per ton of coal sold during the first four years of the period, and $2.00 per ton for coal sold during the last three years of the period. Congress extended the credit for Indian coal for one year in ATRA, reasoning such an extension would "encourage continued mining of coal resources on Indian lands."[44] The credit has subsequently been extended as part of "tax extenders." In addition to extending the credit through 2016, the PATH Act modified the provision, removing the January 1, 2009, placed-in-service requirement and modifying the third party sale requirement. The PATH Act also exempts the credit from the alternative minimum tax (AMT). The credit for the production of Indian coal was extended through 2017 in BBA2018.

Proponents of the Indian coal production tax credit have argued the credit helps compensate Indian coal producers for more restrictive regulatory requirements faced by coal producers on Indian lands, and encourages investment and jobs on Indian lands.[45] Concerns related to the credit include environmental and social considerations related to the use of coal-fired power.[46]

---

[44] U.S. Congress, Joint Committee on Taxation, *General Explanation of Tax Legislation Enacted in the 112th Congress*, committee print, February 2013, JCS-2-13, p. 211.

[45] Adam Lidgett, "Mont. Sens. Float Bill To Extend Indian Coal Tax Credit," *Law360*, April 28, 2017, available at https://www.law360.com/articles/918463/mont-sens-float-bill-to-extend-indian-coal-tax-credit.

[46] Valerie Volcovici, "In Montana's Indian country, tribes take opposite sides on coal," *Reuters*, August 21, 2017, available at https://www.reuters.com/article/us-usa-trump-energy-tribes-insight/in-montanas-indian-country-tribes-take%C2%ACopposite-sides-on-coal-idUSKCN1B10D3 .

In: Energy Policies: Issues and Legislation ISBN: 978-1-53614-485-7
Editor: Doug Higgins

Chapter 6

EPA'S PROPOSAL TO REPEAL THE CLEAN POWER PLAN: BENEFITS AND COSTS*

Kate C. Shouse

SUMMARY

In 2015, when the U.S. Environmental Protection Agency (EPA) promulgated the Clean Power Plan to reduce greenhouse gas emissions from fossil-fueled electric power plants, it concluded that the benefits of reducing emissions would outweigh the costs by a substantial margin under the scenarios analyzed. EPA estimated benefits ranging from $31 billion to $54 billion in 2030 and costs ranging from $5.1 billion to $8.4 billion in 2030, when the rule would be fully implemented.

In proposing to repeal the rule in October 2017, EPA revised the estimates of both its benefits and costs, finding in most cases that the benefits of the proposed repeal would outweigh the costs of the proposed repeal. However, EPA found that under other assumptions, the costs of the proposed repeal would outweigh the benefits of the proposed repeal. This chapter examines the changes in EPA's methodology that led to the revised conclusions about how benefits compare to costs.

---

* This is an edited, reformatted and augmented version of a Congressional Research Service report, R45119, dated February 28, 2018.

Three changes to the benefits estimates of the proposed repeal drive the agency's new conclusions.

- First, it considered only domestic benefits of the Clean Power Plan in its main analysis, excluding benefits that occur outside the United States.
- Second, it used different discount rates, including one higher rate, than the 2015 analysis to state the present value of future climate benefits expected from the Clean Power Plan.
- Third, the analysis reduced some estimates of the human health "co-benefits"— that is, the benefits resulting from pollutant reductions not directly targeted by the Clean Power Plan. Specifically, several scenarios assumed no health benefits below specified thresholds for some air pollutants.

EPA also changed the accounting treatment of demand-side energy efficiency savings. EPA's 2015 analysis treated savings from energy efficiency measures as a negative cost, whereas the 2017 analysis treated them as a benefit. Using the terminology of the proposed repeal, EPA moved energy savings from the cost savings estimate to the forgone benefits estimate. There was no change in the difference between benefits and costs because the benefits and costs increased by the same amount. This change took on more significance in a separate analysis that EPA conducted to analyze the cost savings of the proposed repeal.

EPA based one set of benefit-cost estimates of the proposed repeal on its 2015 power sector modeling, which does not reflect changes that have since occurred in the power sector. EPA based the other set of benefit-cost estimates on more recent power sector projections from the Annual Energy Outlook 2017. The power sector changes subsequent to 2015 are potentially important and include changes in expected electricity demand, expected growth in electricity generation by renewable energy technologies, retirements of older generating units, changes in the prices and availability of different fuels and renewables, and state and federal regulations. While modeling differences render the two sets of estimates incomparable, both sets of estimates show a range of costs exceeding benefits (i.e., net costs), and benefits exceeding costs (i.e., net benefits) of the proposed repeal. EPA stated that it plans to update the power sector modeling and make it available for public comment before it finalizes the proposed repeal. This forthcoming analysis may show the extent to which updated power sector projections may change EPA's benefit-cost estimates.

INTRODUCTION

In October 2017, the U.S. Environmental Protection Agency (EPA) proposed to repeal the Clean Power Plan (CPP),[1] a rule that the agency had finalized in 2015 to limit carbon dioxide (CO_2) emissions from existing fossil-fuel-fired power plants. The CPP has been controversial since its inception, was quickly litigated, and has not gone into effect due to a stay issued by the Supreme Court in February 2016.[2] Disagreements about the CPP have centered on the rule's legal justification, design, and scope.[3] In addition, stakeholders have disagreed about whether the rule's benefits would outweigh the costs.

Under the authority of the Clean Air Act Section 111(d),[4] the CPP established national CO_2 emission rates measured in pounds of CO_2 per megawatt-hour (MWh) of electricity generation for existing fossil-fuel-fired power plants. Based on these national emission rates, EPA calculated state-specific goals and gave states two compliance choices. States could demonstrate compliance through either a "rate-based" approach, under which it would implement measures to achieve a statewide emissions rate goal, or a "mass-based" approach, under which the state would implement measures to achieve a statewide total emissions goal.[5] This formulation, along with other options, allowed states flexibility in how to achieve the standards while minimizing costs or meeting other objectives. The emission reductions achieved would depend on how states chose to comply

---

[1] For the proposed repeal, see EPA, "Repeal of Carbon Pollution Emission Guidelines for Existing Stationary Sources: Electric Utility Generating Units; Proposed Rule," 82 *Federal Register* 48035, October 16, 2017.

[2] For more detail about the legal aspects of EPA's proposed repeal as well as CPP litigation, see CRS Legal Sidebar LSB10016, *EPA Proposes to Repeal the Clean Power Plan*, by Linda Tsang and CRS Report R44480, *Clean Power Plan: Legal Background and Pending Litigation in West Virginia v. EPA*, by Linda Tsang and Alexandra M. Wyatt.

[3] CRS Report R44341, *EPA's Clean Power Plan for Existing Power Plants: Frequently Asked Questions*, by James E. McCarthy et al.

[4] 42 U.S.C. §7411(d).

[5] EPA, "Carbon Pollution Emission Guidelines for Existing Stationary Sources: Electric Utility Generating Units; Final Rule," 80 *Federal Register* 64665, October 23, 2015. See also CRS Report R44145, *EPA's Clean Power Plan: Highlights of the Final Rule*, by Jonathan L. Ramseur and James E. McCarthy.

with the rule and other factors, such as the fuel source used and quantity of electricity generated.

Given that the CPP qualified as an economically significant regulatory action, EPA provided a regulatory analysis of it in 2015 that analyzed the benefits and costs. In this analysis, EPA projected a 32% reduction in total power sector CO_2 emissions nationwide by the time the rule was fully implemented in 2030.[6]

The proposed repeal of the CPP marks a significant change from the previous Administration's stance on its policy and legal issues. EPA based the proposed repeal on a change in the agency's interpretation of the authority it has under Section 111(d) of the Clean Air Act (42 U.S.C. §7411(d)). Under the current Administration, EPA determined that the CPP exceeds the legal authority under Section 111(d) by setting CO_2 emission goals for existing power plants that could "only realistically be" achieved using measures "that cannot be employed to, for, or at a particular source"[7]—that is, a location some characterize as "outside the fence line" of the power plants.

The agency also estimated the benefits and costs of the proposed repeal because it qualified as an economically significant regulatory action. Under the Trump EPA's current legal interpretation— that EPA lacked statutory authority to promulgate the CPP in 2015—the benefits and costs are not germane to the decision about the proposed repeal. The benefit-cost analysis of the proposed repeal is nonetheless consequential because it reveals methodological changes relative to EPA's 2015 analysis. Such changes, which are discussed in this chapter, may influence the way EPA estimates benefits and costs of other proposed regulations.

Members of Congress may have an interest in understanding EPA's analysis of the proposed repeal, in particular how some of the agency's 2017 benefit-cost comparisons differ from those in the 2015 analysis. A

---

[6] EPA projected CO2 emissions in the baseline scenario (i.e., without the CPP) as a 16% reduction by 2030 compared to 2005 levels. According to EPA's analysis, the CO2 reductions achieved in 2030 would represent a 32% reduction in power sector CO2 emissions compared to 2005 levels. EPA, "Carbon Pollution Emission Guidelines for Existing Stationary Sources." See also CRS Report R44341, *EPA's Clean Power Plan for Existing Power Plants: Frequently Asked Questions*, by James E. McCarthy et al.

[7] EPA, "Repeal of Carbon Pollution Emission Guidelines for Existing Stationary Sources."

group of 19 Senators submitted a letter to EPA requesting more information about the benefit-cost analysis for the proposed repeal, including about changes in the way the agency estimated benefits and costs relative to the analysis conducted in 2015 for the final CPP rule.[8] Some of the changes in EPA's 2017 analysis may set a precedent for the way federal agencies account for climate benefits and human health benefits when developing regulations to limit greenhouse gases (GHGs) or conventional pollutants, such as particulate matter.

This chapter summarizes the analysis that EPA conducted of the CPP's impacts as part of its proposed repeal. The chapter also identifies how the economic analysis of the 2017 proposed repeal differs from the approach that EPA used in developing the 2015 CPP final rule. The first section provides background on the regulatory analysis requirements for executive branch agencies, an overview of EPA's 2015 analysis, and a summary of the agency's 2017 analysis. The next section compares the conclusions of EPA's 2015 and 2017 benefit-cost analyses. The chapter then describes three primary changes in the 2017 analysis—namely, the estimation of climate benefits, consideration of human health co-benefits, and the accounting treatment of energy savings impact—and finishes with concluding observations.

BACKGROUND ON REGULATORY ANALYSIS REQUIREMENTS

EPA prepares benefit-cost analyses for significant regulatory actions, such as the CPP, in accordance with Executive Order (E.O.) 12866 and Office of Management and Budget (OMB) Circular A-4.[9] Issued in 1993,

---

[8] U.S. Senator Tom Carper et al., letter to Honorable Scott Pruitt, Administrator, EPA, October 26, 2017, https://www.epw.senate.gov/public/index.cfm/2017/10/senate-democrats-to-epa-show-your-work-on-clean-power-plan-repeal

[9] Federal agencies must comply with a set of regulatory analytical requirements that have been "established incrementally during the last 40 to 50 years through a series of presidential and congressional initiatives, including statutes, executive orders, circulars, and other

E.O. 12866 replaced regulatory analysis directives from prior Administrations with similar but not identical requirements. The analytical principles and requirements for the development and review of federal regulations outlined in E.O. 12866 remain in effect today.[10]

In particular, E.O. 12866 directs federal agencies to examine the benefits and costs of significant regulatory actions and ensure that the benefits justify the costs. It defines significant regulatory actions as rules that meet any one of the following four conditions:

1) It would have an annual impact on the economy of $100 million or more;
2) It would "create a serious inconsistency or otherwise interfere with" other agency actions;
3) It would "materially alter the budget impact of entitlements, grants, user fees, or loan programs or the rights and obligations of recipients thereof;" or
4) It would "raise novel legal or policy issues."[11]

Rules that meet the first condition are considered to be economically significant and are required to have a more detailed regulatory analysis. Whereas E.O. 12866 articulates the principles for regulatory analysis and the overarching requirements, a 2003 guidance document, "OMB Circular A-4," elaborates on what constitutes a "good regulatory analysis" and how to develop one. OMB Circular A-4 defines a good regulatory analysis as one that (1) explicitly states the need for the proposed regulatory action, (2) analyzes alternative approaches to the proposed regulatory option, and (3) assesses the benefits and costs.[12] The circular describes best practices

documents." See CRS Report R41974, *Cost-Benefit and Other Analysis Requirements in the Rulemaking Process*, coordinated by Maeve P. Carey.

[10] Presidents have required agencies to conduct some form of regulatory analysis prior to rule promulgation since 1971. For a summary of the history of the development of regulatory requirements, see CRS Report R41974, *Cost-Benefit and Other Analysis Requirements in the Rulemaking Process*, coordinated by Maeve P. Carey.

[11] E.O. 12866 §3(f), "Regulatory Planning and Review," 58 *Federal Register* 51735, October 4, 1993.

[12] OMB Circular A-4, "Regulatory Analysis," September 17, 2003, see p. 2.

and aims to standardize the measurement and reporting of benefits and costs of economically significant regulatory actions across federal agencies.

Consistent with E.O. 12866, OMB Circular A-4 notes that good regulatory analysis informs consideration of whether benefits of an action "are likely to justify the costs."[13] The "justify" criterion does not require monetized benefits to outweigh monetized costs. E.O. 12866 and Circular A-4 recognize that quantified benefit and cost estimates may not capture all of the anticipated benefits and costs of a regulatory proposal, because it is difficult to quantify some impacts.[14] While Circular A-4 directs agencies to quantify the benefits and costs of economically significant regulations "whenever possible" by applying "sound and defensible values or procedures,"[15] it also directs analysts to identify which non-quantified impacts "are of sufficient importance to justify consideration in the regulatory decision."[16]

Overview of 2015 Regulatory Analysis

When it promulgated the CPP in 2015, EPA determined that the rulemaking was economically significant and therefore prepared a regulatory analysis in accordance with E.O. 12866, OMB Circular A-4, and the agency's *Guidelines for Preparing Economic Analyses*.[17] EPA

---

[13] OMB Circular A-4, p. 2.

[14] E.O. 12866 §(1)(b)(6), "Regulatory Planning and Review."

[15] OMB Circular A-4, p. 27.

[16] OMB Circular A-4, p. 10.

[17] EPA developed *Guidelines for Preparing Economic Analyses* to ensure that the agency's economic analyses inform the policymaking process and meet requirements set forth in relevant executive orders and guidance documents, including E.O. 12866 and OMB Circular A-4. See National Center for Environmental Economics, *Guidelines for Preparing Economic Analyses*, U.S. Environmental Protection Agency, Washington, DC, May 2014, https://www.epa.gov/environmental-economics/guidelines-preparing-economic-analyses. EPA also noted that the regulatory analysis served as the economic assessment required by Clean Air Act Section 317. U.S. Environmental Protection Agency, *Regulatory Impact Analysis for the Clean Power Plan Final Rule*, October 23, 2015, p. 1-4, at https://www.regulations.gov/document?D=EPA-HQ-OAR-2013-0602-37105 (hereinafter, "2015 RIA").

calculated state-specific emission rate goals (pounds CO2 per megawatt hour [MWh]) for the *rate-based* scenario and state-specific emission goals (tons of CO2) for the *mass-based* scenario.[18] EPA characterized these scenarios as illustrative in nature, noting that states could implement the CPP in a variety of ways. EPA emphasized that while the impacts estimated under the two illustrative scenarios were not "definitive," the analysis nonetheless represented "EPA's best assessment of likely impacts of the CPP under a range of approaches that states may adopt."[19]

EPA estimated emission reductions and associated compliance costs under the two scenarios using the Integrated Planning Model (IPM), a detailed model of the U.S. power sector.[20] EPA reported compliance costs as the "projected additional cost of complying with" the CPP in a given year. According to EPA, the compliance cost estimates were based on the

> net change in the annualized cost of capital investment in new generating sources and heat rate improvements at coal-fired steam generating units, the change in the ongoing costs of operating pollution controls, shifts between or amongst various fuels, demand-side energy efficiency measures, and other actions associated with compliance.[21]

These estimates also included the expected costs for monitoring, reporting, and recordkeeping.

EPA expected the CPP to reduce CO_2 emissions as well as non-GHG emissions (sulfur dioxide [SO_2], nitrogen oxides [NO_x], and directly emitted fine particulate matter [PM]). EPA used the IPM model to estimate the reduction in CO_2 emissions and then applied the social cost of carbon

---

[18] For additional explanation and examples of how EPA developed the scenarios, see CRS Report R44341, *EPA's Clean Power Plan for Existing Power Plants: Frequently Asked Questions*, by James E. McCarthy et al.

[19] 2015 RIA, p. ES-3.

[20] Developed by ICF International, IPM projects the economic and environmental impacts of prospective air pollution policies. It projects emission control strategies that the power sector could employ "while meeting energy demand and environmental, transmission, dispatch, and reliability constraints," 2015 RIA, p. 109.

[21] 2015 RIA, p. ES-9.

(SCC) to estimate the economic value of the associated climate benefits.[22] The SCC is an estimate of the monetary value of impacts associated with changes in CO_2 emissions in a given year. It includes net changes in agricultural productivity, property damage from increased flood risk, and changes in energy system costs, such as reduced costs for heating and increased costs for air conditioning.[23]

EPA also estimated the human health benefits of reductions from non-GHG emissions and referred to them as "co-benefits" because the CPP did not directly target those emissions. Specifically, EPA quantified the expected reductions in SO_2 and NO_x emissions—precursor emissions that contribute to the formation of PM and ozone.[24] EPA then monetized the human health benefits expected from reduced exposure to PM and ozone by using a "benefit-per-ton" approach for PM and for ozone. Generally speaking, the benefit-per-ton is an estimate of the average dollar value of human health benefits associated with the reduction of one ton of a pollutant.[25] EPA calculated an average benefit-per-ton estimate in three different regions (Eastern United States, Western United States, and California) for the CPP analysis. These estimates represented the value of

---

[22] For the SCC estimates, see Interagency Working Group (IWG) on Social Cost of Carbon, with participation by Council of Economic Advisers, Council on Environmental Quality, Department of Agriculture, Department of Commerce, Department of Energy, Department of Transportation, Domestic Policy Council, Environmental Protection Agency, National Economic Council, Office of Management and Budget, Office of Science and Technology Policy, and Department of the Treasury, "Technical Support Document: Technical Update of the Social Cost of Carbon for Regulatory Impact Analysis Under Executive Order 12866," May 2013 (revised July 2015), https://obamawhitehouse.archives.gov/sites/default/files/omb/inforeg/scc-tsd-final-july-2015.pdf.

[23] For more information about the SCC, see CRS In Focus IF10625, *Social Costs of Carbon/Greenhouse Gases: Issues for Congress*, by Jane A. Leggett.

[24] In this report, *PM* refers to fine particulate matter. EPA expected the CPP final rule to achieve reductions in directly emitted PM but was not able to quantify those reductions. See 2015 RIA, p. ES-6.

[25] EPA has used this approach in other regulatory analyses. For example, EPA used benefit-per-ton estimates in (1) EPA, Regulatory Impact Analysis: Final Rulemaking for 2017-2025 Light-Duty Vehicle Greenhouse Gas Emission Standards and Corporate Average Fuel Economy Standards, EPA-420-R-12-016, August 2012, https://nepis.epa.gov/Exe/ZyPDF.cgi/P100EZI1.PDF?Dockey=P100EZI1.PDF ; and (2) EPA, Regulatory Impact Analysis of the Cross-State Air Pollution Rule (CSAPR) Update for the 2008 National Ambient Air Quality Standards for Ground-Level Ozone, EPA-452-R-16-004, September 2016, https://www3.epa.gov/ttn/ecas/docs/ria/transport_ria_final-csapr-update_2016-09.pdf

reductions in premature deaths and illnesses, such as non-fatal heart attacks and asthma, associated with exposure to PM and ozone.[26]

PM health effects accounted for most of the monetized health co-benefits estimates. Nearly all of the monetized PM health co-benefit estimates—98%—were from reductions in premature deaths related to PM exposure. Also, most of the value of estimated PM health co-benefits came from SO_2 emission reductions. For example, the monetized health co-benefits for reductions in SO_2 as a precursor to PM—accounted for roughly 80-85% of the estimated dollar value of human health co-benefits in the scenarios analyzed for 2030.[27] Ozone health effects accounted for a smaller share of the monetized health co-benefits—roughly 6-12% of the estimated dollar value of human health co-benefits in the scenarios analyzed for year 2030.[28]

Overview of the 2017 Proposed Repeal

Two years after finalizing the CPP under the Obama Administration, EPA, under the Trump Administration, proposed to repeal the CPP and began the same rulemaking process it had used to promulgate the CPP. Clean Air Act Section 307(d) requires EPA to adhere to the same rulemaking process regardless of whether the agency promulgates, revises, or repeals a rulemaking.[29] The Clean Air Act Section 307(d) requirements most relevant to the regulatory analysis are summarizing the data serving

---

[26] Reductions in non-fatal heart attacks are estimated based on reduced exposure to PM. The asthma-related impacts associated with exposure to PM are based on exacerbation of asthma symptoms in individuals with asthma. The monetized ozone impacts account for emergency room visits for asthma. See 2015 RIA, Table ES-6, for a complete list of the human health impacts considered.

[27] This range is based on a 3% discount rate. The percentage varies by illustrative scenario (rate-based versus mass-based) and benefit-per-ton value used. See 2015 RIA, Table 4-18 and Table 4-21. For similar estimates in year 2025, see 2015 RIA, Figure 4-2, p. 4-36.

[28] This range is based on a 3% discount rate. The percentage varies by illustrative scenario (rate-based versus mass-based) and benefit-per-ton value used. See 2015 RIA, Table 4-18 and Table 4-21. For similar estimates in year 2025, see 2015 RIA, Figure 4-2, p. 4-36.

[29] For more information about procedural requirements, see CRS Report R44341, *EPA's Clean Power Plan for Existing Power Plants: Frequently Asked Questions*, by James E. McCarthy et al., pp. 40-41.

as the basis for the repeal and specifying the methodology used to obtain and analyze the data. While EPA based the proposed repeal on a change in its legal interpretation of Section 111(d) of the Clean Air Act, the agency revisited the 2015 CPP regulatory analysis, because the proposed repeal qualified as an economically significant regulatory action. Therefore, EPA prepared a regulatory analysis in accordance with E.O. 12866, OMB Circular A-4, and the agency's *Guidelines for Preparing Economic Analyses*.[30]

Broadly speaking, the benefits of repealing a rulemaking are the avoided costs that would have been incurred through implementing the rule; the costs of the repeal are the forgone benefits that would have resulted from implementing the rule. EPA defined the benefits of the proposed CPP repeal as the "avoided compliance costs"—that is, the compliance costs that would have been incurred to implement the CPP. EPA also refers to this category as "cost savings." Likewise, EPA defined the costs of the proposed CPP repeal as the forgone reductions in CO_2 and non-GHG emissions and the associated forgone climate benefits and health co-benefits, respectively.

EPA did not conduct new power sector modeling for the 2017 analysis but used two existing power sector projections to estimate two sets of the benefits and costs of the proposed repeal. EPA stated that it "plans to do updated modeling" using IPM, making it available for public comment before it finalizes the proposed repeal.[31]

The first set of benefit-cost estimates was based on the power sector modeling EPA conducted using IPM for the rule in 2015. Specifically, EPA used the 2015 power sector modeling results— estimated compliance costs and estimated reductions in CO_2, SO_2, and NO_x emissions under the mass-based and rate-based scenarios—as the starting point to monetize the avoided compliance costs and forgone benefits of the proposed repeal.

---

[30] EPA also noted that the regulatory analysis serves as the economic assessment required by Clean Air Act Section 317. EPA, *Regulatory Impact Analysis for the Review of the Clean Power Plan: Proposal*, October 2017 (hereinafter "2017 RIA"), p. 27, https://www.regulations.gov/document?D=EPA-HQ-OAR-2017-0355-0110

[31] 2017 RIA, p. 3.

The second set of benefit-cost estimates was based on more recent power sector projections published in the U.S. Energy Information Administration's (EIA) Annual Energy Outlook (AEO) 2017. Specifically, EPA used the AEO 2017 results—estimated compliance costs and estimated reductions in CO_2, SO_2, and NO_x emissions under the CPP mass-based scenario—as the starting point to monetize the avoided compliance costs and forgone benefits of the proposed repeal.

In addition, EPA changed the accounting treatment of cost savings from demand-side energy efficiency measures. EPA counted them as a negative cost in the 2015 analysis. In 2017, EPA moved them to the benefits side of the ledger, counting them as a positive benefit of the CPP. Using the terminology of the proposed repeal, this means that EPA counted the energy efficiency savings—which would not be achieved if EPA were to repeal the CPP—as a forgone benefit.

COMPARING THE 2015 AND 2017 BENEFIT-COST ANALYSES

EPA's 2015 analysis concluded that the monetized benefits of the CPP outweighed the monetized costs. The 2017 analysis presented less favorable benefit-cost comparisons of the CPP. Traditionally, benefit-cost comparisons are shown as estimates of the "net impact," which is the difference between total benefits and total costs. "Net benefits" result when the benefits outweigh the costs, and "net costs" result when the costs outweigh the benefits.[32]

All of the net impact estimates presented in the 2015 analysis showed benefits outweighing the costs, with the difference ranging from $25 billion to $45 billion in 2030.[33] In contrast, the 2017 net impact estimates

[32] While the net impact can provide a rough measure of how the estimated benefits compare to the estimated costs, it does not necessarily determine whether the benefits justify costs. For example, the net impact does not account for potentially important qualitative impacts.

[33] These estimates are net benefits for one year—2030. This range spans estimates across the rate-based and mass-based scenarios and two discount rates (3% and 7%).

ranged from costs outweighing benefits (i.e., net costs of the proposed repeal) to benefits outweighing costs (i.e., net benefits of the proposed repeal).[34] The estimates for 2030 ranged from $28.3 billion in net costs of the proposed repeal to $14 billion in net benefits of the proposed repeal.[35] The 2017 analysis did not explicitly state the agency's view about whether the benefits of the repeal are likely to outweigh the costs of the proposed repeal. Table 1 presents the highest and lowest net impacts reported for each analysis year in the 2015 and 2017 analyses.

Table 1. CPP and Proposed CPP Repeal: Estimated Net Impacts EPA's range of net estimates, based on 2015 power sector modeling, by year (billions of 2011 $)

Year	CPP Final Rule Estimated Range of Net Benefits	Proposed CPP Repeal Estimated Range: (Net Costs) to Net Benefits
2020	$1.0 to $6.7	($3.8) to $2.9
2025	$15 to $27	($18.1) to $6.6
2030	$25 to $45	($28.3) to $14.0

Source: CRS analysis of EPA's 2015 Regulatory Impact Analysis (pp. ES-22 to ES-23) and EPA's 2017 Regulatory Impact Analysis (pp. 71-77).

Notes: "Net Impact" refers to the difference between total benefits and total costs: Net benefits result when the benefits outweigh the costs, and net costs result when the costs outweigh the benefits. Parentheses in the table signify net costs. Ranges in table show the lowest and highest estimated net impacts based on 2015 power sector modeling and as reported in EPA's 2015 and 2017 analyses. Each range spans estimates from the rate-based and mass-based scenarios and two discount rates (3% and 7%).

The benefit-cost estimates derived from the AEO 2017 projections likewise showed a range of net costs and net benefits of the proposed repeal under different assumptions. Half of these benefit-cost comparisons

---

[34] Roughly one-third of the benefit-cost comparisons showed net costs of the proposed repeal, and roughly two-thirds showed net benefits of the proposed repeal (see Tables 4-1 through 4-4 in the 2017 RIA). These estimates in the 2017 RIA were based on the power sector modeling EPA conducted in 2015 with the IPM model. As previously noted, this was one of two sets of benefit-cost estimates. The second set of estimates—based on AEO 2017 power sector projections—are not comparable to the EPA's 2015 estimates due to modeling differences.

[35] These estimates are the net impacts of the proposed repeal for one year—2030. The range spans estimates shown for year 2030 in Tables 4-1 through 4-4 of the 2017 RIA, which were based on the power sector projections used in the 2015 RIA. The range covers the rate-based and mass-based scenarios and two discount rates (3% and 7%).

showed net benefits to repeal the CPP, and the other half showed net costs to repeal. The AEO-based benefit-cost estimates are generally the same order of magnitude as proposed repeal estimates in Table 1, though as previously noted, they are not directly comparable due to modeling differences. The AEO-based estimates of the proposed repeal for 2030 ranged from $30.6 billion in net costs to $14.0 billion in net benefits.[36]

The effect of updated power sector projections on the proposed repeal's benefit-cost estimates is unclear. EPA noted that the benefit-cost estimates based on its 2015 power sector modeling did not reflect changes that have since occurred in the power sector. These changes are potentially important and include changes in expected electricity demand, expected growth in electricity generation by renewable energy technologies, retirements of older generating units, changes in the prices and availability of different fuels and renewables, and state and federal regulations.[37] The AEO 2017 projections showed a shift from higher-emitting sources to lower-emitting sources, suggesting potentially lower CO_2 emission reductions and compliance costs under the CPP relative to 2015 estimates. EPA did not specify whether such trends would alter the agency's conclusions about the benefit-cost comparisons of the proposed repeal. Rather, EPA said it plans to update its power sector modeling and make it available for public comment before it finalizes the proposed repeal.[38]

The AEO 2017 projections also suggested greater SO_2 reductions under the CPP than those EPA estimated in 2015. Specifically, AEO 2017 scenarios showed higher human health co-benefits, due to higher SO_2 reductions, compared to EPA's 2015 estimates.[39] One recent study by

---

[36] These estimates are the net impacts of the proposed repeal for one year—2030. The range spans estimates shown for year 2030 in Tables 7-8 through 7-11 of the 2017 RIA, which were based on AEO 2017 power sector projections. The range covers two discount rates (3% and 7%) and was based on EIA's modeling of the CPP mass-based scenario. EPA also reported net impacts for these scenarios in two other years—2020 and 2025. The 2020 estimated net impacts ranged from $0.5 billion in net costs of the proposed repeal to $0.1 billion in net benefits of the proposed repeal. The 2025 estimated net impacts ranged from $5.0 billion in net costs of the proposed repeal to $14.3 billion in net benefits of the proposed repeal. See 2017 RIA, pp. 126-130.

[37] 2017 RIA, p. 17.

[38] 2017 RIA, p. 3.

[39] The AEO 2017 scenarios showed greater particulate matter benefits due to greater reductions in SO2 emissions under the CPP. Based on AEO 2017 projections, EPA reported that

Resources for the Future analyzed the AEO 2017 projections by fuel type and observed that it projected a greater reduction in coal emissions under the CPP in 2030 relative to what EPA estimated in 2015. The study concluded that it is unclear what mechanisms are "behind the result and to what extent the differences between models used in the 2015 regulatory impact analysis (RIA) projections and the 2017 AEO projections play a role."[40]

What Changed in EPA's 2017 Benefit-Cost Analysis

Compared to the 2015 analysis, EPA changed its analysis of the CPP's benefits and costs in three primary ways. Specifically, EPA:

1) Revised estimates of the SCC by (a) excluding international impacts of U.S. emissions and (b) more strongly discounting the future benefits of CO_2 reductions;
2) Assumed no human health co-benefits below set thresholds of pollution levels in most of the benefit-cost comparisons; and
3) Counted demand-side energy efficiency savings as benefits rather than as offsetting costs.

The first two changes accounted for the differences in the conclusions of EPA's 2015 analysis compared to the 2017 analysis. The third change did not alter the conclusions—whether benefits outweigh the costs—but

relative to the baseline, the CPP would have reduced SO2 emissions by 191,000 and 423,000 short tons in 2025 and 2030, respectively. See 2017 RIA, p. 122. The corresponding SO2 reductions reported in 2015, based on EPA's IPM analysis of the mass-based scenario, were lower: 185,000 and 280,000 short tons in 2025 and 2030, respectively. See 2015 RIA, p. ES-7.

[40] Dallas Burtraw, "Comments to the Maryland Office of the Attorney General and the Maryland General Assembly on the Proposed Repeal of the Clean Power Plan," Resources for the Future, January 11, 2018, http://www.rff.org/files/document/file/RFF-Testimony-Burtraw-Jan2018_3.pdf.

modified the separate estimates of the CPP compliance costs and benefits. This section details all three changes.

New SCC Values Lower Estimates of the Climate Benefits

EPA and other federal agencies have used the SCC to value the climate benefits of CO_2 reductions from rulemakings. Typically presented as dollars per metric ton of CO_2 in a given year, the SCC is an estimate of the monetary value of impacts from a change in CO_2 emissions in a given year. The impacts include net changes in agricultural productivity, property damage from increased flood risk, and changes in energy system costs, such as reduced costs for heating and increased costs for air conditioning. SCC values are calculated using models that translate changes in emissions into economic impacts through a multi-step process.[41] Analysts multiply the SCC estimates for a given year by the estimated CO_2 emissions reduction in that same year to estimate the monetary value of the associated climate benefits.

In 2015, EPA estimated the climate benefits of the CPP using four SCC values developed by an interagency working group (IWG).[42] The IWG SCC estimates, which measured the global value of CO_2 reductions, were \$17, \$53, \$77, and \$160 per metric ton of CO_2 emissions in 2030 (2011\$).[43] The first three values were based on the average SCC from the model runs at discount rates of 5%, 3%, and 2.5%, respectively. The fourth

[41] The resulting SCC estimates do not, however, account for all potentially significant climate change impacts. The integrated assessment models used to estimate the SCC do not include all potentially significant climate change impacts. The National Academies of Sciences, Engineering, and Medicine (NAS) discusses the extent to which current integrated assessment models capture climate change impacts. See NAS, *Valuing Climate Damages: Updating Estimation of the Social Cost of Carbon Dioxide* (Washington, DC: National Academies Press, 2017), Chapter 5, https://www.nap.edu/catalog/24651/valuing-climate-damages-updating-estimation-of-the-social-cost-of.

[42] IWG, "Technical Support Document: Technical Update of the Social Cost of Carbon for Regulatory Impact Analysis Under Executive Order 12866."

[43] The IWG reported SCC estimates in 2007 dollars per metric ton. For the 2015 analysis, EPA adjusted those estimates to (1) short tons using conversion factor of 0.90718474 metric tons in a short ton and (2) 2011 dollars using the GDP Implicit Price Deflator (1.061374). See 2015 RIA, p. 4-7.

value was the 95th percentile of the SCC from the model runs at a 3% discount rate.

In 2017, the Trump Administration withdrew the IWG's SCC estimates and disbanded the IWG as part of a broader executive order, E.O. 13783, which directed federal agencies to review certain federal requirements with respect to domestic energy development.[44]

Discount Rate Basics

Economists use discount rates to compare benefits and costs that occur at different times. Discounting helps answer the question about how much future benefits and costs are worth today. A discount rate adjusts future values based on the observation that people usually prefer a value today compared with the same amount in the future.

The choice of a discount rate has implications for how much one values current consumption over future consumption. Higher discount rates give less present value to benefits or costs that accrue in the future; lower discount rates give more present value.

The following example illustrates this effect. Assume that someone promises to give you $1 billion in 50 years. That sum of money is worth about $228 million today with a 3% discount rate. That is, if you invested $228 million today at a 3% rate and let it compound for 50 years, it would amount to $1 billion. A higher discount rate of 7% would decrease the value today to about $34 million. Alternatively, $34 million invested today at a 7% rate and compounded for 50 years would produce $1 billion.

Recognizing that agencies may need to value climate change impacts in regulatory analysis,[45] E.O. 13783 directed agencies to ensure that new

---

[44] E.O. 13783 required the heads of federal agencies to review, suspend, revise, or rescind "all existing regulations, orders, guidance documents, policies, and any other similar agency actions (collectively, agency actions) that potentially burden the development or use of domestically produced energy resources, with particular attention to oil, natural gas, coal, and nuclear energy resources." See E.O. 13783, "Promoting Energy Independence and Economic Growth," 82 *Federal Register* 16093, March 31, 2017.

[45] Previous judicial decisions have faulted agencies for failing to account for the economic value of climate change impacts in benefit-cost analysis. See CRS Legal Sidebar WSLG1684, *Courts Evaluate How Federal Agencies Put a Price on Carbon*, by Linda Tsang.

estimates are consistent with OMB Circular A4.[46] E.O. 13783 specifically mentioned the Circular A-4 guidance for agencies to focus on domestic benefits and costs and to use discount rates of 3% and 7%.[47] Circular A-4 does not consider issues specific to each regulation, such as those pertaining to climate change, and does not provide guidance that explicitly references the SCC.[48]

EPA Develops Domestic SCC Values and Uses Higher Discount Rate

EPA developed new SCC values—labeling them as "interim values"—based on the direction given in E.O. 13783.[49] EPA used the same models and assumptions as the IWG except with respect to the scope (domestic versus global) and discount rates.[50] For example, EPA and the IWG applied the same projections of population, income, and emissions to the models. EPA, however, used two of the three models to directly calculate domestic SCC estimates. Given that the third model—Dynamic Integrated Climate and Economy (DICE) 2010—generates only global values, EPA used an approximation to adjust those global values to domestic values. Specifically, EPA approximated the domestic SCC as 10% of the global SCC[51] based on a peer-reviewed paper by the DICE 2010 model author.[52] The interim, domestic SCC estimates resulting from all three models are $7

---

[46] E.O. 13783 §5(c).

[47] OMB Circular A-4, p. 15.

[48] Circular A-4 was issued in 2003, before federal agencies developed and used SCC estimates in regulatory analyses. Circular A-4 nonetheless includes a number of provisions relevant to assessing climate change impacts. For example, it directs analysts to "present information on the streams of benefits and costs over time in order to provide a basis for assessing intertemporal distributional consequences, particularly where intergenerational effects are concerned." OMB Circular A-4, p. 14.

[49] 2017 RIA, p. 42.

[50] Whereas the IWG selected four SCC estimates—the average SCC at each of three discount rates (2.5%, 3%, 5%) and the 95th percentile SCC at 3%—for use in regulatory analysis, EPA selected two SCC estimates for the 2017 CPP analysis—the average SCC at two discount rates (3%, 7%). EPA also used the average SCC at 2.5% in a sensitivity analysis. It did not present the 95th percentile value at 3%.

[51] A hypothetical calculation illustrates this approach: A $100 global SCC value would translate to a $10 domestic SCC. See 2017 RIA, p. 162.

[52] Paper reported results from a regionalized version of DICE 2010, known as RICE 2010. See William D. Nordhaus, "Revisiting the Social Cost of Carbon," *Proceedings of the National Academy of Sciences of the United States*, vol. 114, no. 7 (2017), pp. 1518-1523.

and $1 per metric ton of CO_2 emissions in 2030 (2011 dollars) at discount rates of 3% and 7%, respectively.

Considerations for the Scope—Domestic or Global—of the SCC

The United States emits CO_2 and other GHGs to the atmosphere, where the gases become well mixed and contribute to global GHG concentrations, affect the global climate, and have consequences for humans, economies, and natural systems in the United States and in other countries. Most published estimates of the SCC have measured the global impact. Stakeholders, however, have disagreed about whether federal agencies should use domestic or global SCC values.

In 2010, the IWG recommended global values because most GHGs "contribute to damages around the world independent of the country in which they are emitted" and concluded that the SCC "must therefore incorporate the full (global) damages caused by GHG emissions to address the global nature of the problem."[53] EPA agreed with the IWG's rationale in 2015 and observed that the United States operates in a global, interconnected economy and that the potential for spillover effects—particularly in the areas of national security, international trade, and public health—further justified use of global SCC estimates.[54]

Some stakeholders supported this position. In particular, they stated that use of a domestic SCC would understate the benefits to the United States because of spillover effects—that is, climate impacts that occur outside U.S. borders could nonetheless affect the U.S. economy.[55] Others have observed that "U.S. emissions cause the bulk of their damages beyond U.S. borders"[56] and discussed "economic, strategic, ethical, and

---

[53] For the complete discussion of EPA's rationale for using global estimates, see 2015 RIA, p. 4-7.

[54] EPA, *EPA's Responses to Public Comments on the EPA's Carbon Pollution Emission Guidelines for Existing Stationary Sources: Electric Utility Generating Units*, Docket ID EPA-HQ-OAR-2013-0602-37106, Chapter 8, "Economic and Employment Impacts, Sections 8.7-8.9," August 2015, p. 44.

[55] For a summary of the comments recommending global SCC values and EPA's response, see EPA, *EPA's Responses to Public Comments on the EPA's Carbon Pollution Emission Guidelines for Existing Stationary Sources*, pp. 41-45.

[56] William Pizer et al., "Using and Improving the Social Cost of Carbon," *Science*, vol. 346, no. 6214 (December 5, 2014), pp. 1189-1190.

legal justifications"[57] to focus on a global SCC. In short, those recommending global values concluded that "no bright line between domestic and global" climate change impacts exists.[58]

Other stakeholders disagreed with this position, recommending in 2015 that federal agencies use domestic SCC estimates. These stakeholders noted that it would be consistent with OMB Circular A-4 guidance to focus on domestic benefits and costs.[59] Others questioned whether basing U.S. climate regulations on global values would help international climate mitigation efforts.[60] Some stakeholders also criticized the comparison of impacts measured on different scales—global benefits versus domestic costs—and concluded that using a global SCC overstates the benefits of a country-specific rulemaking. That is, the benefit-cost comparison may seem favorable when counting benefits that accrue to non-U.S. populations but would be less favorable when considering only the U.S. benefits.

In the 2017 analysis of the proposed repeal, EPA pointed to direction in E.O. 13873 and OMB Circular A-4 to explain its focus on domestic SCC estimates. Circular A-4 does not, however, prohibit consideration of global values. Rather, A-4 states that analysis should "focus on benefits and costs that accrue to citizens and residents of the United States" and continues to note that when evaluating a "regulation that is likely to have effects beyond the borders of the United States, these effects should be reported separately."[61] EPA presented global estimates as part of a sensitivity analysis in the appendix of the 2017 regulatory analysis. It did not compare those estimates to the avoided costs of the proposed repeal.

---

[57] Peter Howard and Jason Schwartz, "Think Global: International Reciprocity as Justification for a Global Social Cost of Carbon," *Columbia Journal of Environmental Law*, vol. 42 (March 5, 2017), p. 210, http://www.columbiaenvironmentallaw.org/think-global-international-reciprocity-as-justification-for-a-global-social-cost-of-carbon/

[58] EPA, *EPA's Responses to Public Comments on the EPA's Carbon Pollution Emission Guidelines for Existing Stationary Sources*, p. 43.

[59] For a summary of the comments recommending domestic SCC values and EPA's response, see EPA, *EPA's Responses to Public Comments on the EPA's Carbon Pollution Emission Guidelines for Existing Stationary Sources*, pp. 41-45.

[60] See, for example, Ted Gayer and W. Kip Viscusi, "Determining the Proper Scope of Climate Change Policy Benefits in U.S. Regulatory Analyses: Domestic versus Global Approaches," *Review of Environmental Economics and Policy*, vol. 10, issue 2 (July 2016), pp. 245-263, https://academic.oup.com/reep/article-abstract/10/2/245/1753002?redirectedFrom=fulltext

[61] OMB Circular A-4, p. 15.

Experts acknowledge challenges in developing accurate estimates of the domestic SCC, which EPA discussed in its 2017 analysis. In particular, EPA acknowledged the difficulty in capturing the "relevant regional interactions—i.e., how climate change impacts in other regions of the world could affect the United States, through pathways such as global migration, economic destabilization, and political destabilization."[62] The National Academies of Sciences, Engineering, and Medicine (NAS) presented a similar observation, stating that the current models used to estimate the SCC do not fully capture the relevant regional interactions.[63]

Considerations for Discount Rates

OMB Circular A-4 discusses discount rate selection for benefit-cost analysis. OMB based the Circular A-4 discount rate recommendations—3% and 7%—on scenarios in which a regulation would primarily affect either consumer spending or private capital.[64] In one scenario, regulations may directly affect private consumption by, for example, raising consumer prices for goods and services. In another scenario, regulations might displace or alter capital investments in the private sector. The 3% rate, known as the "consumption rate," was based on the rate that the average saver uses to discount future consumption. Specifically, it is the real rate of return on long-term government debt, which averaged about 3% between 1973 and 2003. The 7% rate was based on the opportunity cost of capital—the displaced or forgone investment—and corresponded to the growth rate of federal spending.

Discounting occurs in the last step of the SCC calculation. The models first estimate the climate change impacts that occur over a long time period—to 2100 and beyond—following the release of CO2 and then discount the future value of those impacts to the year of the CO2 emission. Discounting allows for apples-to-apples comparisons of economic impacts

---

[62] 2017 RIA, pp. 45-46.

[63] NAS, *Valuing Climate Damages*, p. 53.

[64] Executive Office of the President, Council of Economic Advisers, "Discounting for Public Policy: Theory and Recent Evidence on the Merits of Updating the Discount Rate," January 2017, https://obamawhitehouse.archives.gov/sites/default/files/page/files/201701_cea_discounting_issue_brief.pdf.

that occur at different times. It is generally standard practice in benefit-cost analysis.[65]

The intergenerational aspect of climate change makes selection of a discount rate particularly challenging when calculating the SCC—in part because it has implications for how much the current generation values the climate change impacts experienced by future generations. The current generation must select a discount rate on behalf of the future generation and without the benefit of input from the future generation. It also raises questions about the extent to which the current generation would account for the future generation's potential preferences, particularly if doing so comes at the expense of the current generation.

While there is no consensus on the appropriate rate to choose in an intergenerational context, it is well understood that higher discount rates result in lower present values and that lower discount rates result in higher present values.[66] In addition, the literature shows that SCC estimates are highly sensitive to discount rate selection.[67]

OMB Circular A-4 also acknowledges the ethical implications involved when comparing benefits and costs that span generations. Circular A-4 recommends that analysts consider conducting sensitivity analysis that applies lower discount rates, noting that estimates of intergenerational rates in the 1990s ranged from 1% to 3%.[68]

OMB Circular A-4 has not been updated since 2003.[69] Under the Obama Administration, the Council of Economic Advisers assessed the Circular A-4 discount rates in light of more recent market data. In January 2017, the council concluded that "the evidence supports lowering these

---

[65] For more information about discounting generally and specifically the considerations for SCC estimation, see CRS In Focus IF10625, *Social Costs of Carbon/Greenhouse Gases: Issues for Congress*, by Jane A. Leggett.

[66] NAS, *Valuing Climate Damages*, p. 161.

[67] For example, see (1) NAS, *Valuing Climate Damages*, p. 161; (2) K. Arrow et al., "Determining Benefits and Costs for Future Generations," *Science*, vol. 341 (July 26, 2013), pp. 349-350; and (3) Michael Greenstone, Elizabeth Kopits, and Ann Wolverton, "Developing a Social Cost of Carbon for U.S. Regulatory Analysis: A Methodology and Interpretation," *Review of Environmental Economics and Policy*, vol. 7, no. 1 (January 2013), pp. 23-46.

[68] OMB Circular A-4, pp. 35-36.

[69] OMB Circular A-4, pp. 33-34.

discount rates."[70] In particular, the council suggested that the 3% rate "should be at most 2 percent" and the 7% rate "should also likely be reduced."[71]

Stakeholders have expressed divergent views on discount rate selection for the SCC. Environmental and public interest groups generally emphasized the intergenerational considerations and recommended that the federal government use lower rates or rates that decline over time.[72] Industry groups have generally recommended higher rates, such as the 7% rate.[73] The published literature largely shows application of lower discount rates in climate change studies. According to the NAS, the majority of climate change impacts studies cited in the Intergovernmental Panel on Climate Change Fifth Assessment Report (2014) used rates of no more than 5%.[74]

In 2010, the IWG concluded that the SCC discount rate should reflect the rate that the average saver uses to discount future consumption (i.e., the consumption rate). The IWG therefore did not apply the 7% rate, which was based on the opportunity cost of capital.[75] As previously noted, the IWG discounted the SCC at rates of 2.5%, 3%, and 5%. In the 2015 CPP analysis, EPA agreed with the IWG's selection of discount rates, characterizing them as consistent with the economics literature and OMB Circular A-4 guidance.[76]

---

[70] Council of Economic Advisers, "Discounting for Public Policy."

[71] Council of Economic Advisers, "Discounting for Public Policy."

[72] For example, see a letter authored by four environmental groups, EPA-HQ-OAR-2013-0602-23545, at https://www.regulations.gov/. For a summary of the comments recommending lower discount rates and EPA's response, see EPA, *EPA's Responses to Public Comments on the EPA's Carbon Pollution Emission Guidelines for Existing Stationary Sources*, pp. 61-65.

[73] For example, see letter submitted in response to the IWG's SCC comment solicitation, OMB-2013-0007-0141 at http://www.regulations.gov. For a summary of the comments recommending higher discount rates and the IWG's response, see IWG, "Response to Comments: Social Cost of Carbon for Regulatory Impact Analysis Under Executive Order 12866," July 2015, https://obamawhitehouse.archives.gov/sites/default/files/omb/inforeg/scc-response-tocomments-final-july-2015.pdf, pp. 21-22.

[74] NAS, *Valuing Climate Damages*, p. 168.

[75] IWG, "Technical Support Document: Social Cost of Carbon for Regulatory Impact Analysis Under Executive Order 12866," February 2010, https://www.epa.gov/sites/production/files/2016-12/documents/scc_tsd_2010.pdf p. 19.

[76] See 2015 RIA, p. 339, and EPA, *EPA's Responses to Public Comments on the EPA's Carbon Pollution Emission Guidelines for Existing Stationary Sources*, pp. 62-64.

Academics have recommended different ways to calculate a discount rate. In particular, the NAS recommended that the federal government develop a new approach to calculate discount rates that would better capture uncertainty over long periods of time. The NAS did not recommend a particular rate but instead focused on the method that federal agencies should use to determine the appropriate rate. Furthermore, the NAS suggested that the OMB Circular A-4 guidance is an insufficient basis for choosing SCC discount rates, because it "does not fully address the issue of discounting over long horizons or the effect of uncertainty on discount rates, both of which directly influence" SCC estimates.[77]

Climate Benefit Estimates Lower in 2017 Analysis

EPA estimated the forgone climate benefits of the proposed repeal using the two interim SCC estimates, which produced notably lower benefit estimates than EPA's 2015 analysis.[78] EPA valued the same tonnage of CO_2 emissions in the 2015 and 2017 analyses but applied lower SCC values in the 2017 analysis. The domestic perspective and use of a 7% rate lowered the SCC estimates and therefore reduced the estimates of climate benefits under the CPP—that is, the forgone climate benefits of the proposed repeal. Table 2 summarizes the climate benefit and forgone climate benefit estimates from the two analyses.

EPA acknowledged the limitations of the interim SCC estimates and conducted a sensitivity analysis with respect to the scope (domestic versus global) and discount rate. Specifically, EPA estimated the forgone climate benefits using global values and an alternative discount rate (2.5%) and presented the results in an appendix. The agency did not, however, directly compare the sensitivity analysis estimates to the avoided compliance costs of the proposed repeal.

---

[77] NAS, *Valuing Climate Damages*, p. 187.

[78] That is, EPA's estimates of the forgone climate benefits under the proposed CPP repeal were notably lower than EPA's estimates of the climate benefits under the CPP final rule.

Table 2. Estimated Climate Impacts under CPP and Proposed CPP Repeal EPA's Estimates for Rate-Based Scenario in 2030 in billions of 2011$

SCC Value	CPP Final Rule Climate benefits in 2015 analysis	Proposed CPP Repeal Forgone climate benefits in 2017 analysis
2.5% global	$29	*$29*
2.5% domestic	—	*$3.9*
3% global	$20	*$20*
3% global (95th percentile)	$61	—
3% domestic	—	$2.74
5% global	$6.4	—
5% domestic	—	—
7% global	—	*$2.5*
7% domestic	—	$0.48

Source: CRS analysis of EPA's 2015 and 2017 Regulatory Impact Analyses.

Notes: EPA used same tonnage of CO2 emissions—415 million short tons in 2030—to calculate the climate benefits in 2015 and the forgone climate benefits in 2017. SCC values are based on the model average unless otherwise noted. The monetized estimates that are italicized were presented in EPA's sensitivity analysis rather than the primary benefit-cost comparison. Categories for which EPA did not report estimates are marked with "—."

Thresholds Reduced Some Human Health Co-Benefit Estimates

EPA's 2017 analysis also diverged from the 2015 analysis by excluding the forgone human health co-benefits from some of the benefit-cost comparisons. While one of the benefit-cost comparisons counted all of the estimated forgone health co-benefits, another benefit-cost comparison completely excluded the forgone health co-benefits. Two of the other benefit-cost comparisons used a threshold that reduced the forgone health co-benefits. In those comparisons, EPA counted only the health co-benefits that exceeded the threshold.

Benefits Estimate for the "Targeted Pollutant" Excluded Health Co-Benefits

One of the comparisons excluded the forgone health co-benefits entirely to focus on the forgone benefits from the "targeted pollutant," CO_2. This presentation appears inconsistent with guidance in OMB's Circular A-4.[79] EPA explained this approach as a way to focus on "the benefits due to reductions in the target pollutant relative to the costs, and whether alternative regulatory designs can achieve reductions in the targeted pollutants and/or the other affected pollutants more cost effectively."[80] EPA further stated that the focus on a targeted pollutant "may be an appropriate way to evaluate this and future regulatory actions" and requested public comment on "the extent to which the EPA should rely on consideration of benefits due to reductions in the target pollutant relative to the costs in the decision-making process."[81]

EPA's rationale for this breakout relates to some stakeholders' concerns about the consideration of co-benefits—typically from reduced exposure to PM—in analyzing regulations targeting other pollutants.[82] For example, stakeholders critical of EPA's 2015 analysis commented that inclusion of the monetized co-benefits made it difficult to understand the impact of the CPP on CO2. Some stakeholders also questioned whether EPA had attributed PM benefits achieved through other regulations to the CPP—that is, "double counted" benefits that would have occurred regardless of the CPP.[83] EPA's 2015 analysis stated that the estimated human health co-benefits were incremental to the baseline that included

---

[79] Circular A-4 states, "Identify the expected undesirable side-effects and ancillary benefits of the proposed regulatory action and the alternatives. These should be added to the direct benefits and costs as appropriate" (p. 3). A-4 also states, "Your analysis should look beyond the direct benefits and direct costs of your rulemaking and consider any important ancillary benefits and countervailing risks" (p. 26).

[80] 2017 RIA p. 11.

[81] EPA, "Repeal of Carbon Pollution Emission Guidelines for Existing Stationary Sources."

[82] For more information about stakeholder views regarding treatment of co-benefits in EPA's economic analyses of Clean Air Act regulations, see CRS Report R44840, *Cost and Benefit Considerations in Clean Air Act Regulations*, by James E. McCarthy and Richard K. Lattanzio.

[83] EPA, *EPA's Responses to Public Comments on the EPA's Carbon Pollution Emission Guidelines for Existing Stationary Sources*, pp. 90-93.

existing regulations of PM and emissions that contribute to the formation of PM.[84]

Consideration of co-benefits and other indirect impacts is typically viewed as a principle of benefit-cost analysis and consistent with federal guidance. OMB Circular A-4 directs agencies to "look beyond the direct benefits and direct costs" of a rulemaking and quantify and monetize co-benefits as well as adverse impacts not already considered in the direct cost estimates.[85] Likewise, EPA's *Guidelines for Preparing Economic Analyses* recommends that the agency's economic analysis "include directly intended effects and associated costs, as well as ancillary (or co-) benefits and costs."[86]

OMB's two most recent reports to Congress on the benefits and costs of federal regulations reported that many of EPA's Clean Air Act analyses monetize co-benefits and count them towards the total benefits estimate. OMB observed that "the large estimated benefits of EPA rules issued pursuant to the Clean Air Act are mostly attributable to the reduction in public exposure to fine particulate matter" and characterized this approach as "consistent with standard accounting practices," noting that this practice "has long been required under OMB Circular A-4."[87]

Two Thresholds Used to Adjust Forgone Health Co-Benefits

The last two benefit-cost comparisons included some but not all of the estimated forgone health co-benefits. EPA applied a threshold to the

---

[84] EPA, *EPA's Responses to Public Comments on the EPA's Carbon Pollution Emission Guidelines for Existing Stationary Sources*, pp. 93-94.

[85] Circular A-4 refers to co-benefits as "ancillary benefits." See OMB Circular A-4, p. 26.

[86] EPA, National Center for Environmental Economics, *Guidelines for Preparing Economic Analyses*, May 2014, p. 11-2, https://www.epa.gov/environmental-economics/guidelines-preparing-economic-analyses.

[87] The quoted language appears in both the 2015 report to Congress and the 2016 report to Congress. The 2015 report is the most recent final version of this annual report. OMB published a draft 2016 report to Congress in December 2016, but it has not been finalized. Both reports can be found at https://www.whitehouse.gov/omb/information-regulatory-affairs/reports/. See (1) OMB, Office of Information and Regulatory Affairs, *2015 Report to Congress on the Benefits and Costs of Federal Regulations and Agency Compliance with the Unfunded Mandates Reform Act*, p. 13; and (2) OMB, Office of Information and Regulatory Affairs, *2016 Draft Report to Congress on the Benefits and Costs of Federal Regulations and Agency Compliance with the Unfunded Mandates Reform Act*, p. 12.

forgone health co-benefits in each of these comparisons, counting only the forgone health co-benefits that exceeded a defined threshold for ambient PM levels. That is, EPA assumed that health co-benefits would equal zero for any PM reductions beyond the defined threshold. While previous EPA analyses have explored uncertainty in benefit estimates relative to benchmark concentrations, the 2017 analysis diverged from past practice by using a threshold to calculate PM mortality impacts.[88]

EPA based one threshold on the current federal air quality standard for PM—12 micrograms per cubic meter (μg/m^3)—and assumed that there were no deaths attributable to PM below it. EPA based the second and lower threshold on the "lowest measured level" (LML) of PM from the two long-term studies it used to estimate deaths related to PM.[89] In this second benefit-cost comparison, EPA assumed that there were no deaths attributable to PM at or below the LML of the Krewski et al. 2009 study (5.8 μg/m^3) and the Lepeule et al. 2012 study (8 μg/m^3).[90]

The PM air quality standard is a higher concentration than the LML and therefore served as a higher threshold. As a result, the PM air quality standard threshold lowered the forgone co-benefit estimates more than the LML threshold did. These thresholds also had implications for the benefit-cost comparisons of the proposed repeal. The comparisons—whether benefits outweigh the costs in various years and at different discount rates—were mostly favorable to the proposed repeal when using the PM

---

[88] For example, EPA's 2015 analysis of the CPP used the lowest measured levels from published studies as a benchmark concentration level to examine the uncertainty of the benefit estimates. The 2015 analysis clarified, however, that EPA did not view this benchmark as a threshold below which benefits fell to zero. See 2015 RIA, p. 4 39.

[89] For more information about the federal air quality standards for PM, see CRS Report R42934, *Air Quality: EPA's 2013 Changes to the Particulate Matter (PM) Standard*, by Robert Esworthy.

[90] See 2017 RIA. EPA used each study to estimate benefits and the results from a range with one "low" estimate and one "high" estimate. EPA used the LML from each study to adjust the high and low forgone benefits. See also (1) D. Krewski et al., "Extended Follow-Up and Spatial Analysis of the American Cancer Society Linking Particulate Air Pollution and Mortality," *Health Effect Institute Research Report*, vol. 140 (2009); and (2) J. Lepeule et al., "Chronic Exposure to Fine Particles and Mortality: An Extended Follow-Up of the Harvard Six Cities Study from 1974 to 2009," *Environmental Health Perspectives*, vol. 120, no. 7 (2012), pp. 965-70.

air quality standard threshold and less favorable when using the LML as a threshold.[91]

EPA explained the use of these thresholds as a way to enhance transparency and "provide some insight into the level of uncertainty in the estimated" PM benefits at lower levels.[92] EPA referenced the agency's 2012 analysis of the rule updating the federal air quality standard for PM, which found greater uncertainty in estimating the "magnitude and significance" of PM-related health risks below the federal standard.[93] Similarly, EPA justified its use of the LML threshold as a way to examine uncertainty in the distribution of PM-related mortality benefits, given that the agency has greater confidence in the estimates that fall within the "bulk of observed" PM concentrations in the two studies.

EPA acknowledged, however, that "scientific evidence provides no clear dividing line" to set a threshold, citing peer-reviewed studies as well as the NAS.[94] One of the cited studies was the agency's 2009 Integrated Science Assessment for Particulate Matter,[95] which evaluated the extensive body of published literature and "concluded that the scientific literature consistently finds that a no-threshold model most adequately portrays the PM-mortality concentration-response relationship."[96]

EPA's 2015 analysis acknowledged lower confidence in benefits that occur from reductions at lower concentrations of PM but did not apply a threshold. At that time, EPA stated that the federal air quality standards are not "risk-free," meaning that there is some risk of adverse health effects from exposure to pollution that meets the federal standard. EPA's 2015 analysis concluded that "the best estimate of benefits includes benefits both above and below the levels of" the federal air quality standard and

---

[91] See 2017 RIA, Tables 4-3 and 4-4. When using the LML threshold, six of the benefit-cost comparisons were favorable to the proposed repeal—that is, they showed net benefits of the proposed repeal—but 10 were not—that is, they showed net costs of the proposed repeal. When using the PM air quality threshold, 11 of the benefit-cost comparisons showed net benefits of the proposed CPP repeal and three showed net costs of the proposed repeal.

[92] 2017 RIA, p. 51.

[93] 2017 RIA, p. 50.

[94] 2017 RIA, p. 51.

[95] EPA, *Integrated Science Assessment for Particulate Matter*, EPA-600-R-08-139F, December 2009, https://cfpub.epa.gov/ncea/isa/recordisplay.cfm?deid=216546

[96] EPA "Clean Power Plan," 80 *Federal Register* 64932, October 23, 2015.

described this practice as consistent with scientific evidence and reviews of the independent Clean Air Scientific Advisory Committee.[97]

EPA's 2017 analysis asked for comment on estimating and reporting PM impacts in this way and described plans for further research. Specifically, EPA intends to "systematically evaluat[e] the uncertainty" associated with the benefit-per-ton approach and compare it to other techniques in the literature in order to better understand the "suitability" of the benefit-per-ton approach for estimating health impacts of changes in pollutants like PM.[98] EPA stated that it intends to conduct, "to the extent feasible," detailed air quality modeling that would, among other things, inform threshold-based analysis of PM benefits.[99]

Change in the Accounting Treatment of Energy Savings Impacts

EPA changed the accounting treatment of demand-side energy efficiency savings. EPA's 2015 analysis treated savings from energy efficiency measures as a negative cost,[100] which reduced the compliance cost estimates. EPA's 2017 analysis treated them as a benefit, citing OMB guidance that described accounting for energy efficiency savings as benefits as a common accounting practice.[101]

Moving the energy savings from the cost side of the ledger to the benefits side simultaneously increased the estimated costs and the estimated benefits by the same amount. In the analysis of the proposed repeal, EPA removed the energy savings (as a negative value) from the total cost estimate, which resulted in a larger cost estimate, and added the energy savings (as a positive value) to the estimated benefits, which resulted in a larger benefits estimate. Using the terminology of the

---

[97] EPA, *EPA's Responses to Public Comments on the EPA's Carbon Pollution Emission Guidelines for Existing Stationary Sources*, p. 102.

[98] 2017 RIA, p. 91.

[99] 2017 RIA, p. 91.

[100] 2015 RIA, p. ES-9.

[101] 2017 RIA, p. 38, referencing OMB, "Guidance Implementing Executive Order 13771, Titled 'Reducing Regulation and Controlling Regulatory Costs,'" 2017, https://www.whitehouse.gov/presidential-actions/presidential-executive-order-reducing-regulation-controlling-regulatory-costs/

proposed repeal, EPA moved energy savings from the cost savings estimate to the forgone benefits estimate. There was no change in the difference between benefits and costs, because the benefits and costs increased by the same amount.

While this accounting change did not change the difference between benefits and costs—that is, the "net impact" of the proposed repeal—it resulted in both higher estimates of the CPP's costs and benefits. This change took on more significance in a cost analysis EPA conducted under E.O. 13771, "Reducing Regulation and Controlling Regulatory Costs."[102] EPA presented the results of its E.O. 13771 analysis in the 2017 regulatory analysis but kept it separate from the benefit-cost comparisons. E.O. 13771 specifies that "any new incremental costs associated with new regulations shall, to the extent permitted by law, be offset by the elimination of existing costs associated with at least two prior regulations," but it does not mention the benefits of regulation.[103]

The analysis that EPA conducted under E.O. 13771 reported the cost savings of the proposed repeal but omitted the forgone benefits. EPA calculated the present value of avoided compliance costs for 2020-2033, using the power sector modeling from EPA's 2015 analysis. EPA did not include the demand-side energy efficiency savings in this cost analysis given that the agency viewed those savings as benefits in the 2017 analysis. Therefore, the cost savings (i.e., CPP compliance costs) analyzed under E.O. 13771 exceeded the compliance costs that EPA considered in the 2015 benefit-cost analysis.

CONCLUSION

Benefit-cost analysis is a core tenet of the regulatory process that informs potentially complex policy decisions. It can strengthen the

---

[102] E.O. 13771 §2(a). E.O. 13771 states that executive departments or agencies must identify at least two regulations for repeal for each new regulation issued. For more information, see CRS Report R44840, *Cost and Benefit Considerations in Clean Air Act Regulations*, by James E. McCarthy and Richard K. Lattanzio.

[103] E.O. 13771 §2(c).

effectiveness of policymaking by providing decisionmakers a consistent framework for evaluating the potential effects of proposed regulatory options. Though it is informative, benefit-cost comparisons—whether the monetized benefits exceed the costs—do not drive a decision on a regulatory proposal. First, there may be important qualitative impacts not captured in the monetized estimates that would inform a determination as to whether the benefits justify the costs of an action. Second, decisionmakers consider an array of factors— including legal considerations, technical feasibility, statutory criteria, and ethical considerations— when developing and finalizing regulatory proposals. For example, EPA based the proposed CPP repeal on a change in the agency's legal interpretation of Section 111(d) of the Clean Air Act and is required to adhere to the rulemaking procedures under Section 307(d) of the act.

Quantifying and monetizing the benefits and costs of a proposed regulation can be a complex task. It may draw upon scientific data from multiple disciplines and involve ethical or policy-based decisions, such as the selection of an intergenerational discount rate. EPA's analyses of the 2015 CPP and the 2017 proposed repeal provide examples of ways that changes in policy priorities and technical information, such as updated information about dynamic factors such as the U.S. power sector, may influence the estimation of benefits and costs. EPA's 2017 benefit-cost analysis reached different conclusions than the 2015 analysis due in large part to the changes in the way it estimated economic impacts. The two changes driving the difference in conclusions were (1) use of new SCC estimates that lowered climate benefit estimates and (2) use of thresholds that reduced human health co-benefit estimates.

Members of Congress have taken divergent views on EPA regulatory policy, in particular the development and promulgation of GHG-related regulations. As EPA considers its proposal to repeal the CPP, Congress, in its oversight role, may consider the extent to which policy priorities influence regulatory decisions. Decisions on regulations such as the CPP have the potential to affect major industries, such as the U.S. energy sector and the health and welfare of U.S. citizens. This report described how EPA's assessment of the benefits and costs of the CPP changed between two Administrations. These changes appear to reflect differences in policy priorities between the Trump and Obama Administrations.

For example, EPA's 2017 analysis used "interim" SCC estimates developed in response to E.O. 13783, which characterized the Obama Administration's IWG SCC estimates as "no longer representative of governmental policy."[104] The interim SCC estimates generally followed the IWG's methodology except with respect to the scope of the estimates—global versus domestic measures of climate change impacts—and discount rates. The domestic perspective and use of a 7% rate lowered the interim SCC estimates, thereby reducing the estimated CPP climate benefits (i.e., the forgone climate benefits of the proposed repeal).

Congress may also consider whether the withdrawal of the IWG SCC estimates—which harmonized the way federal agencies valued GHG impacts in regulatory analysis—will lead to differences across federal agencies in their decisions on GHG-related regulations. EPA's 2017 analysis provided few details about next steps for the SCC and the implications, if any, for other federal analyses. For example, EPA stated that the interim SCC estimates values were developed "for use in regulatory analyses until an improved estimate of the impacts of climate change to the U.S. can be developed based on the best available science and economics" but did not provide a timeline for updates.[105] It is unclear whether other agencies and departments will use the same interim estimates in their own regulatory analyses. To date, at least one other agency—the U.S. Bureau of Land Management (BLM)—has followed a similar approach as the EPA.[106] On the other hand, the Department of Energy continued to use the IWG SCC estimates in one rulemaking, which

---

[104] E.O. 13783 §5(b).

[105] 2017 RIA, p. 42.

[106] BLM used interim, domestic estimates of the social cost of methane to analyze a rulemaking that delayed certain requirements of the 2016 Waste Prevention Rulemaking. See BLM, "Regulatory Impact Analysis for the Final Rule to Suspend or Delay Certain Requirements of the 2016 Waste Prevention Rule," December 2017, https://www.regulations.gov/document?D=BLM-2017-0002-17369.

was published several months after the Administration withdrew those estimates.[107]

In addition, Congress may consider whether EPA's changes to the health co-benefits analysis will set a precedent for future air pollution rulemakings. The human health co-benefits reported in EPA's 2017 analysis reflected different policy determinations than did the 2015 analysis. EPA described the same health impacts literature in the 2015 and 2017 analyses, stating in both analyses that while there is lower confidence in benefits occurring from PM reductions at lower concentrations, the "scientific evidence provides no clear dividing line" to set a threshold.[108] While previous EPA analyses, including the 2015 CPP analysis, explored uncertainty in benefit estimates relative to benchmark concentrations, the 2017 analysis differed by using a threshold to calculate PM mortality impacts. Furthermore, the exclusion of health co-benefits from one of the benefit-cost comparisons departed from the 2015 CPP analysis and federal guidance to consider both direct and ancillary impacts.

---

[107] See, for example, Department of Energy, "Energy Conservation Program: Energy Conservation Standards for Walk-In Cooler and Freezer Refrigeration Systems," 82 *Federal Register* 31808, 31811, July 10, 2017.

[108] 2017 RIA, p. 51; 2015 RIA, p. 4-39.

In: Energy Policies: Issues and Legislation ISBN: 978-1-53614-485-7
Editor: Doug Higgins

Chapter 7

SECTION 201 SAFEGUARDS ON SOLAR PRODUCTS AND WASHING MACHINES*

Vivian C. Jones

INTRODUCTION

On January 23, 2018, President Trump announced that he would impose additional tariffs on imports of *large residential washing machines and solar photovoltaic (PV) cells and modules,* effective February 7, 2018. The President acted based on findings by the *International Trade Commission (ITC)* that increased U.S. imports of these products were a "substantial cause of serious injury" to U.S. manufacturers, as a result of investigations under *Section 201* of the Trade Act of 1974 *(19 U.S.C. §2251,* as amended). When initiating the actions on January 23, the *President said,* "My administration is committed to defending American companies, and they've been very badly hurt from harmful import surges that threaten the livelihood of their workers, of jobs, actually, all over this

* This is an edited, reformatted and augmented version of a Congressional Research Service report, IN10856, dated February 7, 2018.

country." While such actions may benefit some U.S. domestic producers, they could also raise prices for U.S. consumers and domestic industries that use these imports to manufacture downstream products. The Section 201 measures could also increase tensions with various U.S. trading partners.

Section 201 "safeguard" actions provide temporary import relief (e.g., additional tariffs or quotas on imports) in order to facilitate positive adjustment of a U.S. industry to import competition. Safeguard investigations are unlike other trade remedy tools, such as *antidumping (AD)* and *countervailing duty (CVD)* cases that investigate "material injury" (or threat thereof) based on sales of imported products at less than fair value (AD) or that are subsidized by a foreign government or other public entity (CVD). Rather, Section 201 cases investigate import surges of fairly traded goods. Actions taken after Section 201, AD, and CVD investigations will be considered consistent with U.S. international trade obligations if they conform to the relevant *World Trade Organization (WTO) Agreements.* Other WTO members may use these agreements to challenge the United States' actions through the WTO's *dispute settlement process.*

U.S. SOLAR SAFEGUARD ACTION

The Four-Year Safeguard on Solar Cells and Modules

- Places an annual tariff-rate quota (TRQ) of 2.5 gigawatts on solar cells. All imports under that volume level will enter at the normal U.S. (duty-free) tariff rate. Imports above that level will be assessed a 30% tariff, descending by 5% each year (i.e., 30% in 2018, 25% in 2019).

- Places a 30% tariff on solar modules, also descending by 5% each year.
- Applies to all countries except certain developing country WTO members. North American Free Trade Agreement (NAFTA) partners Mexico and Canada are included in the safeguard.
- Instructs the *U.S. Trade Representative (USTR)* to develop procedures to process requests for exclusion of particular products from the safeguard measure within 30 days and to announce regulations in the *Federal Register.*

RESPONSE TO THE SOLAR DECISION

SolarWorld, one of the petitioners in the Section 201 case, expressed hope that the safeguard "will be enough to address the import surge and to rebuild solar manufacturing in the United States." The *United Steelworkers Union,* on behalf of the workers who make silicon, aluminum, steel, and other products used in constructing solar cells and modules, said that the import relief "is important not only to the workers in those sectors, but those who are part of the supply chain," and that the action "is a step forward in drawing a line in the sand that unfair trade will not be allowed."

Other solar industry members expressed concern about the safeguards. The *Solar Energy Industries Association (SEIA),* which represents companies that manufacture, plan, install, and finance solar energy products, estimated that about 23,000 jobs would be lost in 2018 and that the tariffs will "result in the cancellation of billions of dollars in solar investments."

Before the WTO, Taiwan and South Korea have formally requested consultations, the first step in challenging the U.S. action under the *dispute settlement process. China's Ministry of Commerce* announced that China, "with other WTO members, will resolutely defend its legitimate interests."

U.S. Large Residential Washer Safeguard

The Three-Year Safeguard on Large Residential Washers and Parts

- Places an annual TRQ of 1.2 million units on imports of the targeted washers. All imports under that level will enter under the normal U.S. tariff rate (ranging from 1% to 1.4%), plus an additional duty rate of 20% in the first year, decreasing by 2% each subsequent year.
- Requires washers above the quota limit to be assessed at the normal U.S. tariff rate, plus a 50% duty rate in the first year, decreasing by 5% each subsequent year.
- Places an annual TRQ on washer parts of 50,000 units in the first year, ascending by 20,000 units each year. All imports under that volume level will enter at the normal (2.4%) tariff rate. All imports over the level will be assessed a 50% tariff, descending by 5% each year.
- Applies to all countries except for certain developing country WTO members. NAFTA partner Mexico is included in the safeguard, while Canada is exempt.

Response to the Washer Safeguard Decision

Following the Trump Administration announcement, *Whirlpool* reported that it added 200 full-time jobs at its factory in Clyde, Ohio, in anticipation of increased demand following the safeguard decision, and that this would be just the beginning of increased investments and jobs. *Samsung,* one company that produces the targeted washers, said that the announcement "is a great loss for American workers and consumers," and that its new South Carolina plant, opened the week before the safeguard announcement, had already hired more than 600 U.S. workers. These washers, which will be made domestically, will be exempt from the tariffs, but any imported washer parts inputs used in the factory manufacturing process may not be exempt, potentially increasing manufacturing costs and consumer prices. *LG Electronics reportedly said* that the safeguard decision hindered the planning for its new plant in Tennessee.

At the WTO, South Korea has requested *consultations.* A pending *Korean request for retaliation* against the United States in an AD/CVD dispute on washers was also referred to WTO arbitration in mid-January.

In: Energy Policies: Issues and Legislation ISBN: 978-1-53614-485-7
Editor: Doug Higgins

Chapter 8

WINTER FUELS OUTLOOK 2017-2018*

Robert Pirog

SUMMARY

The Energy Information Administration (EIA), in its *Short-Term Energy and Winter Fuels Outlook* (STEWFO) for the 2017-2018 winter heating season, projects that American consumers should expect to see heating expenditures that will be higher than last winter. However, the winter of 2016-2017 was relatively warm. Average expenditures for those heating with natural gas are projected to increase by 12%, while those heating with electricity are projected to see an increase of about 8%. These two fuels serve as the heating source for about 87% of all U.S. household heating. Propane and home heating oil consumers are also projected to see increased costs.

Within the United States, average expenditures projections differences exist with respect to region of the country. Differences in weather conditions and fuel prices contribute to differing regional expenditures.

Economic conditions of relatively high growth and low unemployment suggest that higher consumption levels of all fuels may occur. Increased consumption could lead to higher prices for all winter

* This is an edited, reformatted and augmented version of a Congressional Research Service report, R45042, dated December 5, 2017.

fuels. The STEWFO provides analysis of scenarios covering a warmer, or colder, winter than the base case forecast.

The key federal program designed to assist low-income households is the Low Income Home Energy Assistance Program (LIHEAP). LIHEAP funding beyond the expiration of the Continuing Appropriations Resolution (P.L. 115-56) is uncertain and if reduced could put an additional burden on families qualified for benefits. If the weather in the winter of 2017-2018 is severe, or if energy prices are volatile, the fixed amount of LIHEAP funding will likely result in smaller assistance payments to families, or fewer families receiving payments.

INTRODUCTION

Each year, in October, the Energy Information Administration (EIA) publishes the *Short-Term Energy and Winter Fuels Outlook* (STEWFO).[1] The purpose of the STEWFO is to provide fuel-specific estimates of expected average annual heating expenditures in comparison to previous years. While the STEWFO provides estimates of average annual heating expenses, individual expenses may vary regionally: by energy source utilized, home size, energy efficiency, individual temperature preference, market size, and local weather conditions.

Average annual heating fuels expenditures depend on the price of the fuel used, with natural gas, heating oil, propane, electricity, and wood products constituting the main heating fuels in the United States. Expenditures also depend on the quantity of fuel used, which is partially based on a variety of individual consumer decisions. Weather conditions, measured by heating degree-days, are the other key factor in determining expenditure levels.[2] The National Oceanic and Atmospheric Administration (NOAA) provides winter heating degree-day estimates to the EIA for the STEWFO.

---

[1] The winter heating season runs from October 1 through March 31.

[2] Heating degree-days are defined as the day's average temperature, calculated as the high plus the low temperatures of the day divided by two. If the resulting number is less than 65, subtract the average temperature from 65. The result is the number of heating degree-days on a particular calendar day.

The STEWFO is not a forecast in the statistical sense; it is a projection based on assumed values of key variables. If, for example, the underlying fuel price estimates prove to be incorrect, or weather conditions vary from forecast trends, actual average heating expenditures will reflect those differences.

AVERAGE ANNUAL HEATING FUELS EXPENDITURES

NOAA forecasts a 10.6% increase in heating degree-days for the 2017-2018 heating season compared to 2016-2017 for the United States as a whole. However, the 2016-2017 winter fuel season was warmer than normal. Regional differences in weather, along with regional fuel usage patterns, can cause regional expenditure projections to vary from the U.S. average.

In the Northeast, heating degree-days are expected to increase by 3.5%, while in the West they are expected to increase by 2.9%. Estimated heating degree-days are expected to increase by 23.6% in the South, and increase by 13% in the Midwest.[3]

On average, the EIA projects that U.S. household expenditures on heating fuel for the 2017-2018 heating season will increase for all households irrespective of the heating fuel utilized. The projected increase in expenditures reflects generally higher prices for all fuels, as well as the increased quantity consumed in light of the forecast of colder weather. *Table 1* provides a summary of the percentage changes in the key components of the U.S. average annual heating fuel expenditures by fuel.

Regionally, the EIA expects expenditures on natural gas to increase most in the Midwest at 16.9%, and to increase by 12.1% in the South, 6.9% in the Northeast, and 2.4% in the West. Expenditures on heating oil, used primarily in the Northeast, are expected to increase by 16.8%. Expenditures on propane are expected to increase by 10.6% in the

---

[3] All percentage changes in the STEWFO and this report are calculated on a yearly heating season basis, not on a calendar year basis. The estimated 2017-2018 values are compared to the 2016-2017 values.

Northeast and by 23.5% in the Midwest.[4] Electricity expenditures are expected to increase by 3.4% in the Northeast, 7.7% in the Midwest, 8.6% in the South, and 3.8% in the West.

Table 1. U.S. Average Winter Fuels Projections, Winter 2017-2018 (percentage change)

	Natural Gas	Heating Oil	Propane	Electricity
Consumption	7.7	3.9	N/A	5.1
Price	1.9	12.4	N/A	1.6
Number of Households	0.0	-3.5	-1.3	2.0
Expenditure	9.7	16.8	N/A	6.7

Source: Energy Information Administration, Short-Term Energy and Winter Fuels Outlook, Table WFO1, October 2017.

Notes: Percentage change compares projected changes for the 2017-2018 heating season to 2016-2017 data. The EIA does not provide U.S. average propane data, except number of households.

Due to changing market conditions for primary fuels, the prices assumed in the STEWFO can differ from actual prices. Price volatility in the world oil market can have direct effects on the cost of heating oil and propane. Variations in natural gas prices can directly affect those households heating with that fuel, while indirectly affecting electricity prices. Projected consumption levels are likely to be affected by variations in the weather. If the observed weather is colder, or warmer, than forecast by NOAA, or if consumers change their consumption habits at any given price level, or at any given number of heating degree-days, consumption levels would differ from those projected in the STEWFO.

NATURAL GAS

While the United States is largely self-sufficient in natural gas, it remains part of a North American regional market and participates in

[4] The 2017-2018 *Short-Term Energy and Winter Fuels Outlook* does not include complete data broken down by region for heating oil and propane.

international trade. Most U.S. imports of natural gas come from Canada. Small quantities of liquefied natural gas arrive in the United States from Trinidad. The United States exports natural gas by pipeline to Mexico and Canada, and exports liquefied natural gas to about a dozen nations, with South Korea, Jordan, and Mexico among the largest customers.

Domestic consumers of natural gas include households and commercial customers that largely use natural gas for space heating. Electric power generators, especially those that satisfy peak load demand, use natural gas as a primary fuel to power generators. Industrial consumers use natural gas as a raw material, for example in fertilizer production, and as a heat source in industrial processes. Households, commercial customers, and electric power generators are those consumers whose consumption is most likely to be affected by winter weather conditions.

Table 2 presents average residential consumer natural gas consumption and price data for the winter heating seasons 2013-2014 through projected values for the winter 2017-2018.

Table 2. Average Household Winter Natural Gas Consumption and Price

	2013-2014	2014-2015	2015-2016	2016-2017	2017-2018
Consumption (Mcf)	68.0	64.8	55.8	56.9	61.3
Price ($/Mcf)	$9.97	$9.91	$9.30	$10.11	$10.30

Source: Energy Information Administration, Short-Term Energy and Winter Fuels Outlook, Table WFO1, October 2017.

Notes: Data for winter 2017-2018 are projected, other years are actual values. Mcf is thousand cubic feet.

On a per-household basis, as presented in *Table 2*, winter consumption shows the effects of the last two relatively warm winters compared to the colder two previous winters. The winter of 2017-2018, although projected to generate 10.6% more heating degree-days than last winter, is only an increase compared to the last two warm winters. Approximately 48% of U.S. households heat with natural gas.

Natural gas prices are expected to increase somewhat, an increase of about 2% compared to last year. The observed price-consumption relationship suggests that natural gas demand may be price inelastic in the range presented in *Table 2*. Price inelasticity implies that consumption might not be expected to respond proportionally to changes in price.[6] An inelastic demand relationship is likely because home heating is typically considered to be a necessity by consumers. In addition, existing metering systems do not provide consumers with easy access to real-time price and quantity data. Lack of real-time information could result in consumers making ill-informed decisions with regard to heating usage and expenditures. Income levels and the unemployment rate might also be important in determining natural gas consumption.

The EIA projects a 9.7% average increase in expenditures in households heating with natural gas, driven by a 1.9% increase in natural gas prices, coupled with a 7.7% increase in fuel consumption. Compared to last winter, natural gas heating expenditures are expected to rise by 6.9% in the Northeast, 16.9% in the Midwest, 12.1% in the South, and 2.4% in the West. Average expenditures during the winter of 2017-2018 are expected to be $631 per household, compared to $575 for the winter of 2016-2017. The STEWFO projects, on average, natural gas to be the lowest-cost heating source for the winter of 2017-2018.

On the national level, including all consumer groups, total natural gas consumption rose by about 12% from 2011 through 2016. Over the period, the electric power generating sector's consumption rose by about 40%. Residential and commercial usage showed losses, about -9% and -1% respectively. Industrial demand increased by about 10%, while natural gas consumption in vehicles increased by about 40%, albeit from a small base.[7] The STEWFO projects an increase in total natural gas consumption of about 2% over the period 2016 through 2018.

The STEWFO projects an increase in U.S. natural gas production of about 9% over the period 2016 through 2018. While during the winter of 2016-2017 the United States was a net importer of natural gas, the EIA expects the United States to become a net exporter of natural gas during

[6] Price elasticity of demand is calculated as the percentage change in quantity demanded divided by a given percentage change in price. Demand is considered to be inelastic if the price elasticity calculation yields a value less than one.

[7] Data available at https://www.eia.doe.gov.

the winter of 2017-2018. This shift is due to increased pipeline capacity to Mexico and increased liquefied natural gas export capacity on the Gulf Coast.

Storage levels of natural gas at the beginning of the winter heating season are important because while production is relatively constant over a 12-month period, consumption is subject to seasonal fluctuations. Natural gas in storage is used to smooth out production and consumption differences. The STEWFO projects adequate natural gas in storage for the 2017-2018 heating season. Projected inventories in October 2017, of about 3.8 trillion cubic feet, were above the five-year average.

HEATING OIL

Home heating oil is a middle distillate, derived from the same part of the oil refining process as diesel fuel. As a result, the price of home heating oil is closely related to the price of crude oil as well as the price of diesel fuel. Approximately 4.5% of households in the United States heat with oil, and most of these consumers are in the Northeast, where about 90% of U.S. heating oil consumption takes place.[8]

Table 3. Average Household Heating Oil Consumption and Price

	2013-2014	2014-2015	2015-2016	2016-2017	2016-2017
Consumption (gallons)	607.3	608.1	481.7	517.9	538.1
Price ($/gallon)	$3.88	$3.04	$2.06	$2.41	$2.71

Source: Energy Information Administration, Short-Term Energy and Winter Fuels Outlook, Table WFO1, October 2017.

Notes: Data for 2017-2018 are projected.

The EIA expects average home heating oil expenditures per household to increase by $209, or 16.8%, for the winter of 2017-2018. These dollar and percentage increases are relatively large compared to the cost of heating with other fuels identified in the STEWFO. Heating oil accounts

---

[8] Data available at https://www.eia.doe.gov.

for the highest average total expenditure for any winter fuel, $1,457 per household. The projected heating oil expenditure for winter 2017-2018 is more than double the total expenditure for a household heating with natural gas ($631), and two-thirds higher than a household heating with electricity, $967.

Diesel fuel, a product similar to home heating oil, has been more expensive than gasoline in 2017, averaging about $0.33 per gallon higher. Home heating oil has averaged about the same premium over gasoline during 2017. These price differentials likely result from U.S. refiners' emphasis on gasoline production for the U.S. market, coupled with the relatively high level of world demand for diesel fuel. The United States has exported about 1.4 million barrels per day of diesel fuel over the first eight months of 2017. The costs of gasoline, diesel fuel, and home heating oil are all directly related to the price of crude oil on the world market. The EIA projects the refiner's acquisition cost of crude oil to average $48.75 per barrel in 2017, compared to $40.69 per barrel in 2016.[9]

Two risk factors are key in assessing the heating oil component of the winter fuels outlook. Brent crude oil, the crude oil price most relevant to setting U.S. petroleum product prices, was projected to be $54 per barrel during the winter heating season. However, the spot price of Brent averaged over $62 per barrel during November 2017. Crude oil prices are volatile, and changes in crude oil prices are likely to proportionally affect home heating oil prices.

The second risk factor is the level of inventories in the Northeast. At the beginning of October 2017, distillate inventories in the Northeast totaled 35.5 million barrels, 16.8 million barrels less than at the same time last year. In addition, these inventories were 1.9 million barrels lower than the previous five-year average for the time period. The lower inventory levels are due to high global demand and the disruptive effects of Hurricane Harvey. Prices for home heating oil in the Northeast could rise rapidly if colder-than-expected weather occurs. In that case, the Northeast market might have to turn to more expensive imports. Simultaneous cold

---

[9] Energy Information Administration, *Short-Term Energy and Winter Fuels Outlook,* Table 2, October 2017.

weather in Europe could cause additional upward pressure on price and perhaps compound pressure on low inventories.

PROPANE

Propane provides primary home heating for approximately 5.7 million households in the United States, about 5% of the total households. Propane consumers are projected to experience an increased expenditure of about $211, or 10.6% in the Northeast, while consumers in the Midwest are projected to experience an increase of $276, or 23.5%. These cost increases make propane the second most expensive fuel among those covered in the STEWFO. The number of households heating with propane has remained roughly constant over the past three years. The EIA expects a decline of 1.3% in the number of households using propane in 2017-2018.

Table 4. Average Household Propane Consumption and Price, Northeast

	20213-2014	2014-2015	2015-2016	2016-2017	2017-2018
Consumption (gallons)	745.4	751.5	607.7	651.1	667.6
Price ($/gallon)	$3.56	$3.00	$2.71	$3.06	$3.30

Source: Energy Information Administration, Short-Term Energy and Winter Fuels Outlook, Table WFO1, October 2017.

Notes: Data for 2017-2018 are projected.

Table 5. Average Household Propane Consumption and Price, Midwest

	20213-2014	2014-2015	2015-2016	2016-2017	2017-2018
Consumption (gallons)	868.7	813.2	667.7	679.1	751.7
Price ($/gallon)	$2.61	$1.91	$1.47	$1.73	$1.93

Source: Energy Information Administration, Short-Term Energy and Winter Fuels Outlook, Table WFO1, October 2017.

Notes: Data for 2017-2018 are projected.

For the Northeast in 2016-2017, propane prices were \$3.06 per gallon, while they were \$1.73 in the Midwest, a differential of over 55%. While the consumption levels in the two areas are expected to be, on average, within 90 gallons of each other, the difference in propane prices leads to heating season expenditures of \$1,451 in the Midwest compared to \$2,203 in the Northeast.

Propane is unique compared to other fuels covered in this chapter in the sense that it is a byproduct, and not directly produced itself. The production of gasoline and natural gas both contribute to the supply of propane. As a result, when the supply of those fuels is high, so is the supply of propane.

Many of the same factors that affect natural gas and home heating oil prices also influence expected propane prices. However, in the case of propane, the relationship is indirect because propane is a by-product. Propane prices, unlike the other fuels covered in this chapter, are affected by distance and dispersion of consumers. Because the distribution process usually requires delivery by truck, of relatively small quantities, it tends to be high cost.

ELECTRICITY

Electricity prices are related to natural gas and coal prices as well as the availability of nuclear, hydro, and alternative fuel generating capacity. In addition to natural gas being an input in electricity generation, it is also a direct competitor to electricity as a home heating source. Approximately 40% of U.S. households use electricity as their primary heating source. In the Northeast, electricity use is lowest, at about 15%, while in the South it is highest, at about 63%. Overall growth in the number of households using electricity for space heating is expected to be 2% year-on-year, driven by expansion in the Midwest of over 3%. In comparison, the numbers of propane and heating oil customers are expected to decline, by 1.3% and

3.5%, respectively. The number of households using natural gas for space heating is expected to remain the same for the 2017-2018 heating season.

Table 6. Average Household Electricity Consumption and Price

	2013-2014	2014-2015	2015-2016	2016-2017	2017-2018
Consumption (kwh)	7,981	7,801	7,242	7,227	7,593
Price ($/kwh)	$0.120	$0.123	$0.124	$0.125	$0.127

Source: Energy Information Administration, Short-Term Energy and Winter Fuels Outlook, Table WFO1, October 2017.

Notes: Data for 2017-2018 are projected. Kilowatt hours are kwh.

The EIA projects increasing heating expenditures for households heating with electricity. In the Northeast, expenditures of $1,145, an increase of 3.4%, are projected, and in the Midwest, expenditures of $1,072, an increase of 7.7%, are projected. In the South expenditures of $889 are projected, an increase of 8.6%, while in the West expenditures of $961 are projected, an increase of 3.8%. *Table 6* shows that, while rising, electricity prices have shown less year-on-year volatility than other heating source prices.

Wood

Wood and wood pellets are projected to provide primary home heating for approximately 2.3 million U.S. households in 2017-2018. The EIA also estimates that wood provided a secondary heating source for about 9.3 million households in 2015. Data in the STEWFO show that the number of homes heating primarily with wood and wood products has declined since the winter of 2013-2014, with a year-on-year decline of 1.7% expected during 2017-2018.

The EIA does not track wood prices, and therefore cannot calculate expected expenditure levels for the winter heating season. Wood prices are likely to be highly localized, with relatively large urban/rural differentials, as well as regional differences.

RISK FACTORS

The primary risk factors with respect to STEWFO projections are the weather and energy prices. Total household expenditures on home heating are equal to the price of the fuel times the quantity of fuel consumed. The weather, measured by the number of heating degree-days, largely determines the quantity of fuel used. Conservation, in the form of reduced temperatures inside the home, also can reduce the quantity of fuel consumed, but for a given desired temperature inside the home, heating degree-days are the key factor. The 2017-2018 winter heating season is expected to be colder in all regions of the United States, ranging from 2.9% colder in the West to 23.6% colder in the South. As a result of this key dependence on the weather, the STEWFO includes heating expenditure scenarios estimates based on weather which is 10% warmer, or 10% colder, than the NOAA forecast, identified as the base case, as shown in *Table 7*.

It should be noted that in interpreting *Table 7*, when the weather is warmer, or colder, this affects the quantity demanded of heating fuel, which can translate into a change in the price of the fuel. These "secondary effects" are included in *Table 7.*

The other independent risk factor in determining the expenditure on home heating is the price of the various fuels. These prices are determined by a complex web of related prices, market conditions, exports, expectations, and other economic variables. In a period of relatively strong economic growth, with low unemployment, the key relationship may be that between the level of economic activity, measured by the real growth rate of gross domestic product, and the prices of oil and natural gas.

A higher price of crude oil directly increases the price of home heating oil and propane. Electricity prices are directly affected by natural gas and coal prices. Even in the face of a growing reserve base and increased production, U.S. natural gas prices are expected to increase as shown in *Table 2.*

Table 7. Percentage Change in Home Heating Expenditures (percent)

	Base Case	10% Warmer	10% Colder
Heating Oil	17	5	32
Natural Gas	12	3	19
Propane	18	2	41
Electricity	8	4	12

Source: Energy Information Administration, Short-Term Energy and Winter Fuels Outlook, October 2017.

Notes: The base case compares projected heating expenditures for the 2017-2018 winter heating season to the 2016-2017 winter heating season. The 10% warmer and 10% colder data replace the base case data for 10% fewer, and 10% more heating degree-days.

While crude oil prices are determined in the world market and are also thought to be nonforecastable, the STEWFO includes an assumption of higher oil prices affecting heating oil and propane expenditures as shown in *Table 3* and *Table 4.*

HEATING EXPENDITURE ASSISTANCE

The Low Income Home Energy Assistance Program (LIHEAP) is the primary federal government program to supplement home heating expenditures. LIHEAP is composed of two parts: funding for block grants to states, and emergency contingency funds. LIHEAP funding for FY2018 has been set at $3.39 billion (Continuing Appropriations Resolution, 2018, P.L. 115-56). Prior to the approval of the Continuing Appropriations Resolution, the President's budget request proposed to eliminate funding for the program. Emergency contingency funds have not been appropriated since 2011.[10]

LIHEAP 's ability to provide heating assistance depends upon the level of funding, but also on the cost of energy, which depends upon the prices of the various energy sources; the total consumption of energy for heating,

[10] More detailed analysis of the provisions and funding of LIHEAP can be found in CRS Report RL31865, *LIHEAP: Program and Funding*, by Libby Perl.

which depends on the severity of the weather and energy efficiency; and the number of eligible households.[11]

If the weather in the winter of 2017-2018 is severe, or if energy prices are volatile, the fixed amount of LIHEAP funding will likely result in smaller assistance payments to families, or fewer families receiving payments.

CONCLUSION

The STEWFO projects that U.S. households will generally face increased heating costs during the winter of 2017-2018. Fuel costs are generally expected to be higher than last year and the weather is expected to be colder. Any change in the level of economic activity, or changes in the prices of fuels, coupled with the natural variability in the actual weather experienced during the winter heating season, could change the expenditure projections as shown in *Table 7*.

[11] Ibid., p. 7.

INDEX

A

B

C

D

E

F

G

N

O

P

T

U

V

W